IEE ELECTROMAGNETIC WAVES SERIES 38

Analysis of metallic antennas and scatterers

Other volumes in this series:

Volume 1 **Geometrical theory of diffraction for electromagnetic waves** G. L. James
Volume 2 **Electromagnetic waves and curved structures** L. Lewin, D. C. Chang and E. F. Kuester
Volume 3 **Microwave homodyne systems** R. J. King
Volume 4 **Radio direction-finding** P. J. D. Gething
Volume 5 **ELF communications antennas** M. L. Burrows
Volume 6 **Waveguide tapers, transitions and couplers** F. Sporleder and H. G. Unger
Volume 7 **Reflector antenna analysis and design** P. J. Wood
Volume 8 **Effects of the troposphere on radio communications** M. P. M. Hall
Volume 9 **Schumann resonances in the earth-ionosphere cavity** P. V. Bliokh, A. P. Nikolaenko and Y. F. Flippov
Volume 10 **Aperture antennas and diffraction theory** E. V. Jull
Volume 11 **Adaptive array principles** J. E. Hudson
Volume 12 **Microstrip antenna theory and design** J. R. James, P. S. Hall and C. Wood
Volume 13 **Energy in electromagnetism** H. G. Booker
Volume 14 **Leaky feeders and subsurface radio communications** P. Delogne
Volume 15 **The handbook of antenna design, Volume 1** A. W. Rudge, K. Milne, A. D. Olver, P. Knight (Editors)
Volume 16 **The handbook of antenna design, Volume 2** A. W. Rudge, K. Milne, A. D. Olver, P. Knight (Editors)
Volume 17 **Surveillance radar performance prediction** P. Rohan
Volume 18 **Corrugated horns for microwave antennas** P. J. B. Clarricoats and A. D. Olver
Volume 19 **Microwave antenna theory and design** S. Silver (Editor)
Volume 20 **Advances in radar techniques** J. Clarke (Editor)
Volume 21 **Waveguide handbook** N. Marcuvitz
Volume 22 **Target adaptive matched illumination radar** D. T. Gjessing
Volume 23 **Ferrites at microwave frequencies** A. J. Baden Fuller
Volume 24 **Propagation of short radio waves** D. E. Kerr (Editor)
Volume 25 **Principles of microwave circuits** C. G. Montgomery, R. H. Dicke, E. M. Purcell (Editors)
Volume 26 **Spherical near-field antenna measurements** J. E. Hansen (Editor)
Volume 27 **Electromagnetic radiation from cylindrical structures** J. R. Wait
Volume 28 **Handbook of microstrip antennas** J. R. James and P. S. Hall (Editors)
Volume 29 **Satellite-to-ground radiowave propagation** J. E. Allnutt
Volume 30 **Radiowave propagation** M. P. M. Hall and L. W. Barclay (Editors)
Volume 31 **Ionospheric radio** K. Davies
Volume 32 **Electromagnetic waveguides: theory and application** S. F. Mahmoud
Volume 33 **Radio direction finding and superresolution** P. J. D. Gething
Volume 34 **Electrodynamic theory of superconductors** S.-A. Zhou
Volume 35 **VHF and UHF antennas** Burberry
Volume 36 **Propagation, scattering and dissipation of electromagnetic waves** A. S. Ilyinski, G. Ya. Slepyan and A. Ya. Slepyan
Volume 37 **Geometrical theory of diffraction** V. A. Borovikov and B. Ye. Kinber

Analysis of metallic antennas and scatterers

B D Popović and B M Kolundžija

The Institution of Electrical Engineers

Published by: The Institution of Electrical Engineers, London, United Kingdom

The Institution of Electrical Engineers,
Michael Faraday House,
Six Hills Way, Stevenage,
Herts. SG1 2AY, United Kingdom

British Library Cataloguing in Publication Data

A CIP catalogue record for this book
is available from the British Library

ISBN 0 85296 807 8

Printed in England by Short Run Press Ltd., Exeter

Contents

Preface

Many articles and quite a few book chapters have been written on the numerical analysis of metallic antennas and scatterers in the frequency domain. The major points of these impressive contributions to our knowledge in numerical analysis of electromagnetic fields can be summarised briefly as follows:

- Several types of integral equation have been used for the analysis. For specific types of problem, some of them were claimed to be more convenient than the others
- The equations have been solved numerically mainly by one of the three well known moment-method procedures: point-matching, Galerkin or (in few cases) least-squares
- Subdomain approximation has been adopted in most instances. Only relatively few authors have preferred the entire-domain approach
- For the approximation of metallic surfaces, triangles have been adopted by the majority of authors. A triangle is defined uniquely by three arbitrary points in space so that any system of points over a conducting body can be used to obtain an approximation of the body surface with triangles
- In the approximation of metallic wires, piecewise-cylindrical approximation has been used almost without exception
- For both metallic surfaces and wires, the type of approximation of current varied from the simplest subdomain pulse functions to entire-domain (or almost-entire-domain) polynomials and/or trigonometric functions. Current continuity from one surface or wire element to the other has been observed in most cases
- The difficult problem of wire-to-plate junctions, with or without excitation at the junction, has been treated in considerable detail, but has been solved only for certain classes of junction.

This book describes a relatively simple general unified approach to the analysis of metallic antennas and scatterers. The authors believe that the approach enables the solution to be obtained of a much wider class of problems than with any of the existing methods. In fact, most of the available methods can be regarded to a certain extent as special cases of that elaborated in this book.

Briefly, the approach developed in the book is based on the following principal steps:

(i) On the basis of existing knowledge and many numerical experiments, the electric-field integral equation is adopted as the optimal starting point in the analysis; only for closed bodies is the combined integral equation adopted instead

(ii) On the basis of numerous numerical experiments and comparison with the overall efficiency of the other two procedures, the Galerkin method has been adopted as the optimal procedure for solving the integral equation

(iii) The entire-domain (or almost-entire-domain) approach has been adopted rather than the subdomain approach, because it results in significantly smaller system matrices. For this reason the authors have been advocates of the entire-domain philosophy for over two decades
(iv) Generalised quadrilaterals and their special class, bilinear surfaces (uniquely defined by four arbitrary points in space), are adopted for the approximation of geometry of bodies. These surface elements can model practically any surface more efficiently and with fewer and larger elements than the triangles
(v) A unique treatment is introduced of noncylindrical forms that are always associated with cylindrical segments of wires (such as various types of wire ends or changes in a wire's radius). This has been achieved by introducing generalised wires, and in particular their special forms the truncated cones (with conical or flat rings or discs as important degenerate forms), instead of cylindrical wire segments. In addition, generalised wires are considered as special cases of generalised quadrilaterals, so that plates and wires are modelled with essentially the same tools
(vi) The entire-domain approximation for current is expressed in the same co-ordinates as the surface elements over which it is assumed to exist. A procedure is introduced for converting any set of basis functions into a new set which automatically satisfies current continuity at all interconnections; this helps to reduce the number of unknowns. The polynomial expansion is adopted as the principal expansion. It has been used by the authors and their colleagues for many years, and has been found to be both the simplest and the most flexible
(vii) Finally, a procedure for treating the wire-to-plate-junction problem is proposed which does not require any additional types of surface elements or current expansions (so-called attachment modes), as do all previous methods.

A few useful simple new techniques will also be found in the book. Of these, quite interesting may be a theorem which enables the excitation located on the plate of a wire-to-plate junction to be (approximately) transferred to the wire. This simplifies considerably the analysis of wire-to-plate junctions with excitation. However, the procedure of highly economical and very accurate numerical integration of the integrals appearing in the solution may be found to be even more versatile.

The authors hope that engineers, scientists and graduate students interested in the analysis of electrically small and medium-sized metallic antennas and scatterers will find in this book a very powerful tool that circumvents most difficulties encountered in other available methods.

This book could hardly have been written without the many discussions the authors had during the last decades with their colleagues at the Department of Electrical Engineering of the University of Belgrade. The ideas compressed in this book are certainly partly theirs, although they may not have spelled them out explicitly. In that respect, the authors wish to express their gratitude particularly to Prof. M. B. Dragović and Prof. A. R. Djordjević.

A significant part of the basic ideas and a part of the examples have been published during the last few years in the *Proceedings of the Institution of Electrical Engineers, Part H*. The authors are grateful for the permission to use this material.

B.D.P.
B.M.K.
Belgrade
May 1994

Chapter 1

Introduction

1.1 Basic concepts and definitions

This book is aimed at presenting a general computer-oriented method for the analysis of electrically small and medium-sized metallic objects. The objects are assumed to be situated in a lossless homogeneous dielectric medium in an arbitrary incident monochromatic electromagnetic field. The term 'electrically small objects' refers to objects the maximum dimension of which is much smaller than the free-space wavelength, and 'electrically medium-sized objects' to those the maximum dimension of which does not exceed a few wavelengths. The method is not intended for the analysis of electrically large objects, i.e. those the maximum dimension of which is many wavelengths. However, the limitation is due not to the method itself, but rather to the speed and storage capabilities of available digital computers.

The principal aim of the analysis is to determine the surface-current distribution on the objects. Once this has been determined with sufficient accuracy, all the quantities of interest, such as the scattered (or radiated) field, the near field, the impedance of generators driving the system etc. can be obtained with relative ease.

The first step in the analysis is the description of the structure geometry. Arbitrary metallic structures can be considered as composed of appropriately interconnected flat or curved plates. Such a representation of metallic wires of electrically small radius is also possible, but obviously not at all practical. Therefore thin wires are not decomposed into simpler substructures, except in approximating curved wires by a sequence of straight wire segments or in approximating wire ends and abrupt changes in a wire's radius by various types of caps namely by flat or conical rings.

In principle, the representation of arbitrary curved metallic structures can be strict or approximate. For strict representation of structures by plates, it is necessary to introduce curved plates of arbitrary shape. It will be demonstrated in detail in a later chapter how this can be done. However, such a strict representation of metallic-structure geometry is both theoretically quite complicated and in practice mostly unnecessary. Except for structures whose properties depend critically on their geometry (e.g. some resonant structures), for reasonably accurate analysis of the system it is sufficient to approximate the exact shape of the structure by a convenient system of flat or curved plates. Such a strict or approximate representation of the actual structure geometry is referred to briefly as the modelling of geometry or 'geometrical modelling' of the structure.

For arbitrary curved wires, which in the general case may be of variable cross-section, the same reasoning applies. Except for some resonant wires, it is possible to approximate (model) them as a sequence of cylindrical segments, without introducing appreciable systematic error in the results of the system analysis.

Once the geometrical modelling of a structure has been performed, the next step in the structure analysis is determination of the currents in this approximate structure, for a given incident (impressed) field distribution. Except for some electrically small structures, the exact current distribution cannot be found. Instead, an approximate solution for the current is searched for, most often in the form of a series of known functions (so-called 'expansion functions' or 'basis functions') with coefficients to be determined.

The current approximation depends to some extent on the type of geometrical modelling of the structure which is adopted. If the modelling is such that electrically small (usually flat) patches are used, it is possible to assume that the surface-current-density vector is constant over such a patch, or at the most that it is a slowly varying function of co-ordinates (e.g. a linear function). This approximation of current distribution is known as the 'subdomain approximation', and the corresponding basis functions are termed 'subdomain basis functions'.

If, on the other hand, the geometrical modelling is such that electrically medium-sized surface elements are used to approximate the structure surface, two basic approaches are possible. In one approach, for current-distribution approximation we can subdivide the surface elements into sufficiently small subelements (without changing the shape of the original surface element) so that the subdomain approximation over such subelements is feasible. Another possibility is to determine current approximation over the whole surface element, which is referred to as the 'entire-domain current approximation'. The corresponding basis functions are known as the 'entire-domain basis functions'. If such a surface element is subdivided for current approximation into few relatively large subelements, the current approximation is said to be of 'almost-entire-domain' type, and the basis functions used in this case are termed 'almost-entire-domain basis functions'.

The subdomain current approximation is conceptually and computationally simpler than the entire-domain or almost-entire-domain approximation. However, for electrically medium-sized structures the number of unknowns to be determined becomes quite large, resulting in various numerical problems. The entire-domain and almost-entire-domain approximations tend to be conceptually and computationally more complicated, but enable much larger structures to be analysed on a given computer than the subdomain approximation. The subdomain approximation has also other disadvantages, e.g. there are serious problems in computing accurately the near field owing to the discontinuous nature of the subdomain current expansion or its derivative. As a final remark at this point, let us mention that the surface elements can always be chosen to be so small that the entire-domain and almost-entire-domain approximations become virtually subdomain approximations.

The final aim in the system analysis is the determination of surface-current distribution in the adopted geometrical model of the structure and with the adopted basis functions. This can be done by solving numerically any of a number of integral or integro-differential equations that can be formulated for the problem. The mathematical basis for such a solution, the method of moments, has been known for a long time [1, 2]. It is certain that it has been used many times for solving individual problems, without realisation that the same principle can be applied to the solution of any electromagnetic problem formulated by means of integral or integro-differential equations (in fact, by means of any so-called linear-operator equation). For example, a basic method of moments was used for the analysis of linear antennas in the early 1950s [3, 4].

The application of the method of moments for solving electromagnetic-field problems was apparently first systematically investigated by Harrington [5, 6]. Since the publication of his monograph [6] in 1968, the numerical solution of most problems in electromagnetism has been based on the moment method. Strangely enough, the original paper submitted to the *IEEE Transactions* was rejected because reviewers did not think the problem could be solved in that way (Harrington, R. F.: Private communication, 1992). A more extensive version of the paper was too long for the *IEEE Transactions on Antennas and Propagation*, but by then the *IEEE Proceedings* had expressed a desire to publish the paper [5]. It was published there, and became, without any doubt, the most general and probably the most frequently cited paper on numerical analysis of electromagnetic field problems.

1.2 Brief review of the method of moments

Although the method of moments is very well known, to make this book more or less self-contained it may be worthwhile to explain the basic concepts of the method using a simple example. In this manner we shall also be able to explain the meaning of different expressions to be used in later Chapters of the book.

Consider the electrostatic example of an isolated charged conducting body situated in a vacuum (Figure 1.1). Let the potential of the body be V_0. We wish to determine the charge distribution on the surface of the body.

Since the electric field inside the body is zero, by virtue of the equivalence theorem (e.g. Reference 7) we can remove the conducting body and consider the actual charge distribution in a vacuum. The surface-charge density $\sigma = f$ (f will be used instead of σ because this example serves for the explanation of the general procedure, valid for any unknown) on the former surface of the body must satisfy the integral equation

$$\frac{1}{4\pi\epsilon_0} \oint_S \frac{f(\mathbf{r}')\,\mathrm{d}S}{R} = V_0 \tag{1.1}$$

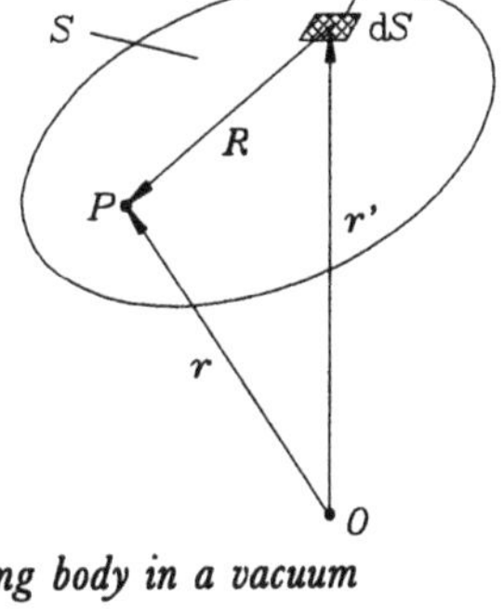

Figure 1.1 Charged conducting body in a vacuum

where

$$R = |\boldsymbol{r} - \boldsymbol{r}'| \tag{1.2}$$

is the distance between the field point (defined by the position vector $\boldsymbol{r}$) and the source point (defined by $\boldsymbol{r}'$). Of course, in this case both $\boldsymbol{r}$ and $\boldsymbol{r}'$ are on the surface S of the body.

The integral equation (eqn. 1.1) can be written for *all* points on S, i.e. for any $\boldsymbol{r}$. Since the integral on the left-hand side symbolises an infinite sum, the equation actually represents an infinite number of linear algebraic equations for an infinite number of unknown surface-charge-density values $f(\boldsymbol{r}')$ on the whole surface of the body. Except in some special cases (e.g. a spherical or ellipsoidal surface), this equation does not have a solution in closed form. However, it can in principle be solved with any accuracy by means of the following procedure.

Assume that the surface-charge density on the body can be represented approximately in the form of a finite sum

$$f(\boldsymbol{r}') = \sum_{i=1}^{N} A_i f_i(\boldsymbol{r}') \tag{1.3}$$

where A_i are unknown coefficients to be determined, and $f_i(\boldsymbol{r}')$ are known basis functions. Substituting eqn. 1.3 into eqn. 1.1 we obtain

$$\frac{1}{4\pi\epsilon_0} \sum_{i=1}^{N} A_i \oint_S \frac{f_i(\boldsymbol{r}')\,\mathrm{d}S}{R} = V_0 \tag{1.4}$$

Since this equation also has to be satisfied at all points of S, it is now an overdetermined system of equations in the unknowns A_i, $i = 1, 2, \ldots, N$, and cannot be solved exactly except in a few simple cases. However, several procedures are available for approximate solution of such equations. To simplify notation, let us rewrite eqn. 1.4 in the form

$$\sum_{i=1}^{N} A_i F_i(\boldsymbol{r}) = g(\boldsymbol{r}) \tag{1.5}$$

The meaning of the functions $F_i(\boldsymbol{r})$ is obvious, and of course in our case $g(\boldsymbol{r}) = V_0$.

The simplest way of solving this equation approximately is to adopt N points on the surface of the body and to require that the equation be satisfied at these points. In this manner a system of N linear equations in the N unknowns, A_i, is obtained, from which these unknowns can easily be determined. This approach to solving such equations is known as the 'point-matching method'. We shall see that, in fact, this is a special case of a much more general method, the method of moments.

The method of moments for solving eqn. 1.5 consists basically of the following. We first adopt a set of n convenient functions $w_j(\boldsymbol{r}), j = 1, 2, \ldots, N$, known as the 'weighting functions' or 'testing functions'. To obtain a set of N linear equations in A_i, we must perform N independent operations over the left- and right-hand sides of eqn. 1.5 which eliminate the variable $\boldsymbol{r}$. Obviously, there is an infinite number of such operations. In the terminology of the method of moments, such operations are known as the 'inner products', and are usually written as $\langle w_j(\boldsymbol{r}), x_i(\boldsymbol{r})\rangle$, where $x_i(\boldsymbol{r})$ stands for either the left-hand or right-hand side of eqn. 1.5. Thus, by performing the operation of an inner product on eqn. 1.5 with all the N testing functions w_i, we obtain the following system of linear equations resulting from the application of the method of moments to eqn. 1.5:

$$\sum_{i=1}^{N} A_i \langle w_j(\boldsymbol{r}), F_i(\boldsymbol{r})\rangle = \langle w_j(\boldsymbol{r}), g(\boldsymbol{r})\rangle \qquad j = 1, 2, \ldots, N \tag{1.6}$$

The usual inner product in the method of moments applied to problems of the type considered is another integration over S (this time with respect to the field co-ordinates $\boldsymbol{r}$). The reason for this is simple. As explained, there is no unique solution of eqn. 1.4 for A_i. We wish that any solution we might obtain be as close to the actual solution as possible. An inner product is therefore desirable that reflects as much of the problem geometry as possible, and this is certainly obtained by another integration over S.

With the inner product adopted, the general method of moments can be formulated in a number of special forms, obtained by choosing different sets of testing functions. We shall refer to these forms as different 'test procedures'. For example, if $w_j(\boldsymbol{r}) = f_j(\boldsymbol{r})$, the method is known as the 'Galerkin method'. If $w_j(\boldsymbol{r}) = \delta(\boldsymbol{r})$ (the Dirac delta function), the previously mentioned point-matching method (or procedure) results. Finally, if $w_j(\boldsymbol{r}) = F_j(\boldsymbol{r})$, the so-called 'least-squares method' (or procedure) is obtained.

Another concept of interest to us can also be understood from the simple electrostatic example considered. Not only the surface of the body but also all points inside the conducting body are at the potential V_0. Therefore we can map the surface S, point by point, to obtain a new surface S' completely inside the interior of the body, and move all the field points in eqn. 1.14 to that new surface. The rest of the solution procedure obviously remains unchanged. Since the conditions for the potential (that it be V_0) are now satisfied not at the body surface, i.e. on the boundary, but instead at a surface inside the body, they are referred to as the 'boundary conditions in the extended sense' or briefly the 'extended boundary conditions'.

Finally, let us write the above equations in the usual notation of the method of moments. First, we use the formal operator notation and instead of eqn. 1.1 write

$$\mathrm{L}f = g \tag{1.7}$$

The operator L in this particular case is defined to be such that the left-hand sides of eqns. 1.1 and 1.7 mean exactly the same. In other cases it can be completely different, but the basic procedure remains the same. The symbol g is used instead of V_0 for convenience, to stress that this can be any function of co-ordinates. Now approximating f by the series expansion in eqn. 1.3, we obtain

$$\sum_{i=1}^{N} A_i \mathrm{L} f_i = g \tag{1.8}$$

We next adopt appropriate weighting functions w_j, $j = 1, 2, \ldots, N$, and perform the inner-product operation on this equation with all the weighting functions in turn, to obtain

$$\sum_{i=1}^{N} A_i \langle w_j, \mathrm{L} f_i \rangle = \langle w_j, g \rangle, j = 1, 2, \ldots, N \tag{1.9}$$

This is the same system of equations as in eqn. 1.6. It can be further simplified (concerning the notation) by introducing matrices. We write

$$\boldsymbol{F}\boldsymbol{A} = \boldsymbol{G} \tag{1.10}$$

where $\boldsymbol{F} = [\langle w_j, Lf_i \rangle]_{N \times N}$, $\boldsymbol{A} = [A_i]_{N \times 1}$, $\boldsymbol{G} = [\langle w_j, g \rangle]_{N \times 1}$ and in $\boldsymbol{F}$ the subscript j stands for rows and i for columns.

There are two more terms from the theory of linear spaces used in the method of moments. The permitted set of functions f_i is termed the 'domain' of the operator L. The set of functions resulting from the operator acting on all the functions in the domain of the operator is termed the 'range' of the operator L.

The outline of the method of moments presented in this Section is a very brief one. Mathematicians would certainly find in it formal deficiencies and lack of rigour. It is believed, nevertheless, that it is sufficient for understanding most of the presentations in the book. The interested reader is referred for further details on the method of moments and a more rigorous treatment to References 1, 2, 6 and 8.

1.3 Basic integral equations for solving electromagnetic-field problems

This book deals only with metallic antennas and scatterers, which are considered as perfect electric conductors. Consider a perfectly conducting body (which may consist of several parts) situated in a homogeneous dielectric medium of parameters ϵ and μ (Figure 1.2). Let the incident (impressed) electromagnetic field ($\boldsymbol{E}_i$, $\boldsymbol{H}_i$) be time-harmonic, of angular frequency ω.

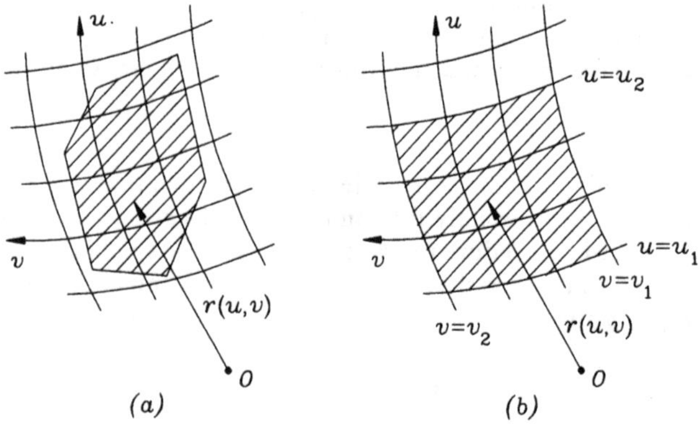

Figure 1.2 Perfectly conducting body in homogeneous dielectric medium excited by incident electromagnetic field

Since there is no field inside the (perfectly conducting) body, the body can be removed, and the actual currents can be considered to exist in an homogeneous medium, on the former surface of the body. (This is again an application of the general-equivalence theorem.) According to the uniqueness theorem on solutions of Maxwell's equations, for a unique solution it suffices that, on the entire body surface, at points just outside it, the boundary condition for either the tangential electric field or the tangential magnetic field be satisfied. Thus, the solution of either of the following two equations which is true on the entire surface S, at points just outside it, is the required solution:

$$\boldsymbol{E}_{\text{tang. due to currents and charges}} + \boldsymbol{E}_{i\,\text{tang}} = 0 \tag{1.11}$$

$$\boldsymbol{H}_{\text{tang. due to currents and charges}} + \boldsymbol{H}_{i\,\text{tang}} = \boldsymbol{J}_s \times \boldsymbol{n} \tag{1.12}$$

In these equations tang. denotes tangential components to S, $\boldsymbol{J}_s$ is the surface-current density vector, and $\boldsymbol{n}$ is the unit vector normal to S and directed into the dielectric.

The electric- and magnetic-field vectors due to currents and charges on the body can be expressed in terms of the magnetic vector potential $\boldsymbol{A}$ and the electric scalar potential V in the usual way,

$$\boldsymbol{E}_{\text{tang. due to currents and charges}} = (-j\omega\boldsymbol{A} - \text{grad } V)_{\text{tang.}} \tag{1.13}$$

$$\boldsymbol{H}_{\text{tang. due to currents and charges}} = (\text{curl}\,\boldsymbol{A})_{\text{tang.}}/\mu \tag{1.14}$$

The potentials can be expressed in terms of the surface-current density $\boldsymbol{J}_s$ and surface-charge density σ on the body as (see Figure 1.2)

$$\boldsymbol{A}(\boldsymbol{r}) = \mu \oint_S \boldsymbol{J}_s(\boldsymbol{r}')g(\boldsymbol{r}, \boldsymbol{r}')\,\mathrm{d}S \tag{1.15}$$

$$V(\boldsymbol{r}) = \frac{1}{\epsilon} \oint_S \sigma(\boldsymbol{r}')g(\boldsymbol{r}, \boldsymbol{r}')\,\mathrm{d}S \tag{1.16}$$

where

$$g(\boldsymbol{r}, \boldsymbol{r}') = \exp(-j\beta R)/(4\pi R) \quad R = |\boldsymbol{r} - \boldsymbol{r}'| \tag{1.17}$$

is the Green function for the homogeneous medium in which the conducting body is situated, and $\beta = \omega\sqrt{(\epsilon\mu)}$ is the phase coefficient.

The operators grad and curl on the right-hand sides of eqns. 1.13 and 1.14 have to be performed at the field point, i.e. over the field co-ordinates $\boldsymbol{r}$. The grad operator acts therefore on the Green function in the integral on the right-hand side of eqn. 1.16 only, and the curl operator on the vector $\boldsymbol{J}_s(\boldsymbol{r}')g(\boldsymbol{r}, \boldsymbol{r}')$ where $\boldsymbol{J}_s(\boldsymbol{r}')$ should be considered as a constant vector with respect to the unprimed co-ordinates. Since, for any vector function $\boldsymbol{F}$ and any scalar function f,

$$\text{curl}\,(f\boldsymbol{F}) = (\text{grad } f) \times \boldsymbol{F} + f\,\text{curl } \boldsymbol{F} \tag{1.18}$$

for a constant vector $\boldsymbol{F}$ we have that curl $(f\boldsymbol{F}) = -\boldsymbol{F} \times \text{grad} f$. Therefore the boundary-condition equations (eqns. 1.13 and 1.14) finally become

$$\left[-j\omega\mu \oint_S \boldsymbol{J}_s(\boldsymbol{r}')g(\boldsymbol{r}, \boldsymbol{r}')\,\mathrm{d}S - \frac{1}{\epsilon}\oint_S \sigma(\boldsymbol{r}')\,\mathrm{grad}\{g(\boldsymbol{r}, \boldsymbol{r}')\}\,\mathrm{d}S\right]_{\text{tang.}} = -\boldsymbol{E}_{i\,\text{tang.}}(\boldsymbol{r}) \tag{1.19}$$

$$\left[-\oint_S \boldsymbol{J}_s(\boldsymbol{r}') \times \mathrm{grad}\{_g(\boldsymbol{r}, \boldsymbol{r}')\}\,\mathrm{d}S\right]_{\text{tang.}} - \boldsymbol{J}_s(\boldsymbol{r}) \times \boldsymbol{n} = -\boldsymbol{H}_{i\,\text{tang.}}(\boldsymbol{r}) \tag{1.20}$$

The unknown to be determined in the latter integral equation is only the surface-current-density vector $\boldsymbol{J}_s(\boldsymbol{r})$. In eqn. 1.19 we have as unknowns both $\boldsymbol{J}_s(\boldsymbol{r})$ and $\sigma(\boldsymbol{r})$. Note, however, that the surface-current density and the surface-charge density are interconnected by the continuity equation for surface currents and charges,

$$\left.\begin{aligned} &\mathrm{div}_s\boldsymbol{J}_s = -j\omega\sigma \\ &\text{i.e. } \sigma = j(\mathrm{div}_s\boldsymbol{J}_s)/\omega \end{aligned}\right\} \tag{1.21}$$

where div_s is the surface divergence. Therefore eqn. 1.19 is also an equation (of integro-differential type) in the unknown surface-current-density vector.

Eqn. 1.19 is known as the 'electric-field integral equation' (EFIE), and eqn. 1.20 as the 'magnetic-field integral equation' (MFIE). There are also other integral or integro-differential equations that can be formulated, but they can all be derived from these two basic types.

The theorem on uniqueness of solutions of Maxwell's equations states that a solution is unique also if the electric-field boundary condition is satisfied on one part of the surface S, and the magnetic-field boundary condition on the rest. Thus it is possible to use eqn. 1.19 on one part of S, and eqn. 1.20 on the rest.

The strict formulation of the theorem on uniqueness of solutions of Maxwell's equations requires that the system has losses, however small. If we neglect losses (as we usually do), but also if losses are very small, and the structure has internal resonances, serious numerical problems are naturally encountered. A more detailed discussion of this problem will be presented in Chapter 6.

It is not difficult to understand that eqns. 1.19 and 1.20 are linear-operator equations. Hence they can be solved approximately by means of the method of moments as explained in Section 1.2.

1.4 Short review of existing methods for the analysis of metallic antennas and scatterers

We have seen that the method of moments offers several numerical procedures for solving electromagnetic problems. However, the actual solution procedure is not at all simple, and, in addition, requires different philosophies in different cases. In the numerical solution of the problems considered in this book, we have first to approximate the actual system geometry and then, considering this simplified structure, to approximate current distribution on it. In both of these steps there is also a multitude of possible approaches, so that many combinations can be adopted in solving an electromagnetic-field problem.

As a result, a large amount of material has been published on various methods of analysing metallic antennas and scatterers. It is therefore virtually impossible to present here any complete review of the work done and to give to the many authors the proper credit they deserve. This book, rather, is aimed at presenting a novel unified method for the analysis of metallic antennas and scatterers. This Section is therefore intended to help the reader in his personal evaluation of the method presented in later Chapters. A brief overview, but as far as possible a systematic one, of the work done that is connected with the subject of the book is therefore presented in what follows.

1.4.1 Wire antennas and scatterers

Of all metallic structures, antennas and scatterers assembled from arbitrarily interconnected straight thin-wire segments are simplest for numerical analysis. Since, in addition, the methods for analysis of more complicated structures have many common points with those for the analysis of wire antennas, it seemed natural to begin this brief review with a description of the basic contributions in developing the methods for thin-wire antenna analysis.

The first rigorous approach to the analysis of wire antennas and scatterers appears to have been due to Pocklington [9]. The integral equation he formulated, named after him, is an integro-differential equation of the EFIE type in unknown current distribution along cylindrical wires. It cannot be solved analytically (which is true for all integral equations of the electromagnetic field), so that for a long time its value was purely academic.

A different integral equation for thin wires, and an iterative solution for it, was proposed by Hallén [10]. (The Hallén equation belongs to the so-called vector-potential-equation group, which can be derived from the EFIE equation for thin wires.) In the early 1950s Nomura and Hatta [3] and Storm [4] proposed, in essence, the moment method for the analysis of cylindrical antennas by means of the Hallén equation, but the solution was very much restricted, because classical computational means were not suitable for the numerical solution of integral equations. In 1956, R. W. P. King in his monumental monograph on linear antennas [11] elaborated in great detail a method similar to the iterative method of Hallén. In 1965, Mei [12] generalized the Hallén equation to arbitrary wire structures. This was probably the end of the era of basically analytical methods of considering the thin-wire-antenna problem. At about the same time, wide use of digital computers made practicable extremely powerful numerical methods for thin-wire-antenna analysis.

The first numerical solution of a wire-structure seems to be due to Richmond [13] (1965). He used the subdomain rectangular basis function and several types of entire-domain basis function to solve the Pocklington equation for a cylindrical wire scatterer by point matching. Harrington [5, 6] in 1967 and 1968 introduced the moment method for solving electromagnetic-field problems in general, and the thin-wire-antenna problem in particular. In considering thin cylindrical antennas, he adopted rectangular and triangular subdomain basis functions for solving the two-potential equation by means of point-matching and by the Galerkin method. Many papers which followed were based on the method of moments, and differed in

- the equation adopted for the analysis
- the method adopted for solving the equation (point-matching, Galerkin etc.)
- the type of basis function (subdomain, entire-domain or almost-entire-domain)
- the treatment of wire junctions and ends
- the approximation of the wire-antenna excitation.

Let us mention some of the contributions which illustrate this diversity of approaches.

Richmond [14] proposed piecewise-sinusoidal approximation of current along thin wire antennas. This approximation allows the electric-field vector to be evaluated in closed form [15].

Analysis of cylindrical symmetrical and asymmetrical antennas by means of Hallén's equation and point-matching was proposed in 1970 and 1971 in References 16 and 17, using entire-domain polynomial approximation for a current of arbitrary degree which automatically satisfies the condition that current goes to zero at the dipole ends. The entire-domain approach and the flexible polynomial current approximation have since been preferred by Popović and a number of other authors, because it required far fewer unknowns and could therefore be used for the analysis of much more complex problems on a given computer than using the subdomain current approximation.

The problem of wire junctions (points at which two or more wire segments meet) appears to have first been solved by Taylor [18] in 1969 and Taylor *et al.* [19] in 1970. Chao *et al.* solved the problem of junctions in 1971 [20], by stipulating the first Kirchhoff's law at a junction in a specific manner and using a technique of overlapping triangular basic functions. Silvester and Chan in 1972 used the entire-domain polynomial approximation for current for the analysis of wire antennas by means of the Galerkin method [21], and in 1973 proposed a general method for the analysis of wire antennas assembled from arbitrarily interconnected straight-wire segments by essentially the same method [22]. In 1972 Tsai [23] introduced a TEM magnetic-current frill to approximate coaxial-line excitation of monopole antennas above a ground plane. In 1973 the same problem was solved approximately by introducing the so-called finite-size-belt voltage generator on the antenna surface [24], and entire-domain polynomial approximation of current was applied to solving the problem of cylindrical antenna with concentrated loadings [25]. Continuity of current was forced across loadings, and continuity of current first derivative was in addition required across loadings of zero magnitude.

In 1973, Thiele [26] treated the problem of many classes of wire antennas without junctions. In 1975, extensive analysis of various thin-wire-antenna solutions was presented by Imbriale [27], Miller and Deadrick [28] and Poggio and Miller [29], including the junction problem.

Although Mei formulated the generalised Hallén equation for arbitrary wire structures in 1965, a practical general approach to the wire-antenna problem starting from the Hallén equation was not attempted owing to quite complicated dual integrals that had to be evaluated numerically. However, in 1976

Kominami and Rokushima [30] proposed a method for analytical evaluation of one of these integrals and analysed some special cases of wire antennas using entire-domain-polynomial approximation of current. In 1979, a general method was formulated by Djordjević *et al.* [31] for analysis of wire antennas assembled from straight segments based on the two-potential equation, point-matching and entire-domain-polynomial approximation of currents, with the TEM magnetic-current frill as the excitation mechanism. In 1982 Nakano [32] offered a similar approach to that of Kominami and Rokushima. In the same year, a monograph [33] was published on numerical analysis of wire antennas by point matching and entire-domain polynomial approximation of current for solving the two-potential equation, and on synthesis (design) of wire antennas by means of numerical-optimisation procedures.

In 1984 a generalisation of the concept of equivalent radius of thin cylindrical antennas was proposed [34], indicating that an infinite number of approximately equivalent antennas exists for any antenna of noncircular cross-section, and even for antennas of circular cross-section. As a special case, it enabled narrow strip antennas printed on thin dielectric substrate to be analysed by the thin-wire methods of analysis. In 1984, the same authors extended the concept of complementary electromagnetic structures [35], which made it possible for the analysis of narrow cylindrical slot antennas, printed on thin dielectric substrate or cut in a sheet of nonzero thickness, to be performed by thin-wire-analysis methods.

In 1986, a general program was completed for the analysis of general wire structures assembled from straight segments [36] based on the extended generalised Hallén equation of Kominami and Rokushima, to result in a monograph on the computer-aided design (CAD) of wire antennas in 1991 [37]. In 1990, Djordjević *et al.* [38] offered a program for the analysis of wire antennas based on the two-potential equation and polynomial current approximation.

Although a number of important wire-antenna problems have not been solved to complete satisfaction (e.g. that of precise modelling of wire junctions and ends), it can be seen from this brief overview that in the 1980s the interest in thin-wire-antenna analysis was modest. This was natural, since much more attention had to be devoted to more complex problems. For example, numerical methods were proposed for the analysis of a combination of wires and metallic bodies, for wire-grid modelling of metallic bodies and for the analysis of metallic and dielectric bodies in an incident electromagnetic field. A brief review of the first group of works follows.

1.4.2 Surface antennas and scatterers

Antennas are often made of wires, but metallic surfaces of various shapes in antennas are of very frequent occurrence. These surfaces can be parts of the antenna itself, or of metallic objects in the vicinity of the antenna. For example, an antenna can be constructed from metallic sheets, plates, cones etc. or a wire antenna can be mounted on a vehicle or an aircraft. Surfaces of metallic bodies in the vicinity of the antenna must be included in the analysis if accurate prediction of the antenna properties is desired.

The analysis of arbitrary metallic surfaces in an incident electromagnetic field is much more difficult than that of thin wires, for several reasons, of which the following seem to be the most important. First, it is often complicated to approximate such surfaces, i.e. to determine an appropriate geometrical model of a surface. Secondly, the surface-current-density vector has two unknown components at all points of the surface. Thirdly, the approximation of each surface-current component over the geometrical model adopted is usually not at all simple. Fourthly, the treatment of surface interconnections and edges, as well as of interconnections of wires and surfaces, requires considerable attention and ingenuity. Fifthly, the excitation of surface structures, if localised in a small domain, also poses serious difficulties. Although certain techniques for thin-wire antenna analysis can be modified so that they can also be applied to surfaces, it is obvious that many additional problems need to be solved and that radically different techniques are necessary for the electromagnetic analysis of metallic surfaces.

The most natural application of our knowledge of thin-wire-antenna analysis to surface metallic structures is to model such surfaces by a wire grid. Although wire-grid modelling has often been used to predict the radiation field, it can frequently give only a vague idea of the quantities depending on the near field, such as the antenna impedance. The purpose of this book is not to consider wire-grid models of metallic surfaces, and therefore in the brief survey of the development of the methods for the electromagnetic analysis of metallic-surface structures that follows wire-grid modelling will not be mentioned.

The earliest work on surface antennas and scatterers seems to be that of Oshiro [39] in 1965. This is another example of the application of the method of moments (although not in its strict form) before it was formulated as a general method. Oshiro considered closed surfaces and divided them into curvilinear rectangles. The surface-current-density vector on every rectangle was assumed to have two unknown constant components, which he determined by an approximate solution of the magnetic-field integral equation and point matching. The excitation was assumed to be independent of the currents over the body.

The first application of the strict moment method to surface-type antennas and scatterers was proposed in 1969 by Mautz and Harrington [40]. They analysed bodies of revolution by means of the Galerkin method, with subdomain basis functions along the body generatrix and entire-domain basis functions around the body. In 1972 Knepp and Goldhirsh [41] described a general method based on the MFIE and subdivision of the body into rather sophisticated surface elements (so-called biquadratic spline surfaces), which enabled them to model a body as complicated as a helicopter. Albertsen *et al.* [42] in 1974 proposed a method for the analysis of structures composed of both metallic surfaces and wires. For the wires they formulated the EFIE, and for the plates the MFIE. The problem of wire-to-surface junctions they solved by introducing a so-called attachment mode, i.e. a specific surface element. Essentially their method was used as the basis for the well known NEC code [43] (Burke and Poggio, 1977).

In 1975 Wang *et al.* [44] proposed the analysis of surface antennas by means of the EFIE, rectangular surface elements and sinusoidal current distribution along the surface-current components. In 1978 Richmond, Pozar and Newman proposed [45], and Newman and Pozar applied [46] an improved sinusoidal distribution over rectangular patches for the analysis of composite wire/surface structures. Among other examples, they solved the problem of a vertical monopole mounted in the middle of a horizontal rectangular metallic surface, driven at the junction by a delta-function generator.

Glisson [47] in 1978 and Glisson and Wilton [48] in 1980 used rectangular surface elements with piecewise-linear current distribution and solved the EFIE, but later, together with Rao, adopted triangular surface elements as more convenient for the approximation of curved surfaces [49]. It is worth mentioning that, in triangular surface elements, they considered three components of the surface-current-density vector instead of two, in spite of the fact that triangular surface elements were planar.

It appears that the first successful attempt to solve a surface problem by entire-domain basis functions was proposed in 1981 in Reference 50. A wire dipole with a rectangular reflector was analysed starting from the EFIE, with polynomial approximation for the surface-current-density vector in two perpendicular directions and using point-matching. The current continuity at the reflector-plate edges was satisfied by stipulating additional constraints. A year later, in Reference 51, a corner reflector was considered instead. It was typical of that approach that, when compared with the subdomain basis functions, very few unknowns per square of the wavelength (typically about 16 per current component) were needed to obtain an accurate solution.

In the period 1981–1983, Schaeffer, Medgyesi–Mitschang, Eftimiu and Putnam, in coauthor pairs, solved several problems of wires and surfaces of revolution and of translation by means of the Galerkin method [52–56], in some instances using entire-domain basis functions [54]. They also considered the junctions problem and that of proper modelling of wire ends.

From 1982 to 1984, Newman, Pozar, Tulyathan and Alexandropulos, again in coauthor pairs, offered a surface-patch model for polygonal plates [57], analysis of a monopole mounted near an edge or a vertex [58, 59], and presented the results for quite large and complicated real structures such as the Boeing 747 and *Concorde* [60].

A solution for a triangular plate antenna without and with passive elements by means of entire-domain polynomial basis functions, EFIE and point matching was proposed in 1986 and 1987 in References 61 and 62. The solution was found not to be very stable, so in these papers an overdetermined system of equations was solved in the least-squares sense, with added line currents along the plate edges to take care of edge-current singularity and to stabilise the solution. Nevertheless, the number of unknowns required was found to be rather large for an entire-domain approximation of current.

A combined subdomain/entire-domain solution for scattering on 2-dimensional objects was proposed in 1988 by Bornholdt and Medgyesi–Mitschang [63]. They used the entire-domain approximation (that was assumed to go to zero at the edges) further from the edges, and subdomain approximation for current distribution near the edges.

Finally, more recently the authors of this book have developed an entire-domain solution of metallic antennas and scatterers based on the EFIE [64–67]. The Galerkin solving procedure of the EFIE and basis functions that automatically satisfy the current-continuity equation at interconnections and ends of wire, or surface elements have been adopted, to circumvent the previously mentioned deficiency of the point-matching method and to reduce the number of unknowns. Basically that approach is also adopted in this book.

1.4.3 Notes on numerical analysis of thin-wire and surface antennas and scatterers

In the brief review on the development of methods for the analysis of thin-wire and surface antennas and scatterers no attempt was made either to describe in more detail various approaches, or to point out the problems that were encountered. However, there are several problems common to most of the methods for numerical analysis of thin-wire and surface antennas and scatterers, and in fact to most of the methods for numerical solution of electromagnetic-field problems in general. Here we aim to point out some of them, with only brief comments. For a more detailed discussion the reader is referred to Poggio and Miller [29], Miller and Deadrick [28], Mittra and Klein [68], Sarkar [69, 70], Sarkar *et al.* [71, 72] and Djordjević and Sarkar [73].

Probably the most acute is the problem of a large number of unknowns, with the associated problems of large computer storage requirements, large matrices, stability and convergence of the results with the increasing order of approximation, if larger structures are considered. It is evident that, for a given structure, entire-domain or almost-entire-domain basis functions require far fewer unknowns than do subdomain basis functions. Although the latter have the advantage of simpler programming, it is nevertheless surprising that entire-domain basis functions have been used so rarely. It has been found by the present authors that almost all electromagnetic problems can be solved by means of the entire-domain type of approximation. With appropriate attention and effort, this was found to result in very stable and accurate solutions and a manyfold reduction in computing time needed for the analysis.

With perfectly conducting structures, which are considered in this book, there is the problem of resonances. Although some procedures have been suggested for circumventing such difficulties, it appears that these procedures are relatively complicated [74]. Therefore, a simple procedure which could alleviate the problem of resonances would be of significant practical value.

As mentioned, three principal methods (or procedures) have been used for the analysis of wire/surface structures: point-matching, Galerkin and least-squares. It was found that the least-squares solution, although theoretically most attractive, appears to be very sensitive to end and edge effects and is therefore not very pleasing from the practical point of view. Of the other two, the Galerkin method is more complicated than point-matching, mainly because it requires an additional integration. However, with a moderate additional analytical effort these dual integrals can be integrated numerically very efficiently. In contrast to the unique results obtained by the Galerkin method, the results obtained by point-matching always depend to a certain degree on the adopted positions of matching points. Although this may not be of great importance in practice, particularly for thin-wire antennas, this should nevertheless be considered as a shortcoming of the point-matching method in general.

Finally, there is obviously no best choice of either subdomain or entire-domain basis functions. However, to obtain a rapid and accurate solution it is important that appropriate attention be paid to that choice. For example, basis functions defined in such a manner that current continuity is satisfied automatically at interconnections, ends and edges of wires and surfaces should be considered superior to those that do not have this property.

1.5 Conclusions

This brief overview of numerical solutions of thin-wire and surface metallic antennas and scatterers presented in this chapter can be summarised as follows:

- Geometrical modelling of wires is performed by right circular cylinders. Current distribution along the wires is approximated in several ways, mostly by subdomain basis functions. Entire-domain basis functions are used only rarely
- Geometrical modelling of surfaces is carried out mostly by approximating the surfaces by a large number of small rectangles or triangles. The current-density vector over a rectangle or triangle is assumed to be constant, linear or sinusoidal
- The problem of interconnections of plates and wires is solved by introducing so-called attachment modes
- The excitation of wire/surface structures can be continuous (e.g. by a plane wave), or more or less concentrated (point generator along wires, TEM magnetic-current frill or belt generator)
- Several types of integal or integro-differential equation are available for the formulation of the problem: EFIE, MFIE, Hallén-type equations, and some combinations of these
- These equations are solved numerically, in most instances by the point-matching procedure or the Galerkin method. The method of least-squares is used only rarely.

In this book a procedure is developed for numerical analysis of metallic antennas and scatterers that has the following principal features:

(i) Modelling of wires is carried out by generalised wires, which as special cases include cylindrical wires, conical wires, various forms of wire ends etc. Entire-domain basis functions are adopted for approximation of current along the wires, which satisfy automatically the continuity equation at generalised wire ends and interconnections

(ii) Generalised curved quadrilaterals and triangles are introduced for geometrical modelling of surfaces. Entire-domain basis functions are adopted for approximation of surface currents over generalised surface elements which automatically satisfy boundary conditions on the surface element edges and interconnections

(iii) Special attachment modes from one case to another of interconnections of wires and surfaces are avoided by modelling surfaces in such regions by three or four curvilinear rectangles. A concentrated excitation on a plate, at the interconnection with a wire (e.g. due to a magnetic-current frill), is transferred to the wire, thereby also avoiding the attachment mode required by a concentrated excitation on the plate

(iv) The excitation of the structure is arbitrary

(v) The electric-field integral equation (EFIE) is adopted for the analysis. Only in the analysis of closed structures is a different integral equation, the so-called combined-field integral equation (CFIE), used instead

(vi) The Galerkin method is adopted for the solution of the integral equation. Multiple numerical integration required by the Galerkin method, which represents probably its principal shortcoming, is performed so efficiently that this deficiency has almost been removed.

Chapter 2

Modelling of geometry of metallic antennas and scatterers

2.1 Introduction

As mentioned in Chapter 1, the first step in the analysis of metallic antennas and scatterers is to describe mathematically the geometry of the structure, i.e. model the structure geometry.

As a rule, geometrical modelling is not a simple task. Frequently the structure geometry cannot be represented exactly in any simple manner, so that a decision must be made how to approximate it properly. However, even when it is possible to represent it exactly, approximate modelling may be more convenient, because it frequently permits more efficient analysis. Of course, exact modelling and the corresponding analysis, when possible, are very important as a check of accuracy of the analysis based on approximate modelling.

The second step, i.e. determination of the current distribution over the structure surfaces, may (and in the approach adopted in this book will) require the definition of two co-ordinate systems. One co-ordinate system, which we may term the 'global co-ordinate system', serves for the definition of the structure as a whole, for the description of near-field points, far-field points etc. The other, auxiliary co-ordinate system (or possibly systems) is defined over part of the exact or approximate structure surface and is used for the description of current distribution on that part of the surface. We shall refer to this co-ordinate system as the 'local co-ordinate system'. In some cases, the global and local co-ordinate systems may coincide.

The description of the (approximate) structure geometry and of the current distribution may, in principle, be carried out in two local co-ordinate systems. Obviously, it is advantageous to adopt the same co-ordinate system for the description of both, and this is what one always tends to do. Therefore, the approximation of the structure geometry and, in particular, the choice of local co-ordinate systems with respect to which the approximate structure is described, is implicitly tightly connected with the determination of current distribution in a later step.

The level of complexity in determining the current distribution depends to a great extent on the choice of the global- and, in particular, of the local co-ordinate system. Therefore it is of great practical interest to find an 'optimal' co-ordinate system, in which the determination of current distribution is as simple as possible. We shall see that, unfortunately, this is usually quite complicated. Nevertheless, it is obvious that special care should be exercised in modelling the structure geometry, because it has a pronounced influence on the whole solution process.

As an example, consider a metallic spherical scatterer. It is completely described by its radius. It appears that the spherical co-ordinate system with the origin at the sphere centre would be the natural and the optimum choice for both the global and local co-ordinate systems. In that case the surface-current-density vector is decomposed into its θ and ϕ components. However, for $\theta = 0$ and $\theta = \pi$ the unit vector $\boldsymbol{i}_\phi$ is undefined, so that at these two points the surface-current-density vector cannot be defined. Therefore, this local co-ordinate system, although convenient for describing the scatterer geometry, cannot be used to describe the surface-current distribution over the whole scatterer. Some local co-ordinate systems which do not have this deficiency will be described later in this Chapter.

To permit approximate (or possibly exact) modelling of the geometry of a wide variety of structures, flexible surface elements are needed which are easy to define mathematically. To that aim, in Section 2.2 generalised quadrilaterals (including as special cases generalised wires, generalised triangles, bodies of revolution and bodies of translation), convenient for the determination of current distribution, are defined. Modelling by such quadrilaterals is illustrated by the example of a spherical scatterer. The spline approximation is applied next to generalised quadrilaterals, with the emphasis on right truncated cones and bilinear surfaces. Finally, special care is exercised in modelling the geometry of wire-to-wire, plate-to-plate and wire-to-plate junctions.

2.2 Generalised quadrilaterals

2.2.1 Definition of generalised quadrilaterals

It is both more natural and simpler to expand the surface currents in a 2-dimensional than in a 3-dimensional co-ordinate system. Modelling of geometry should therefore preferably be carried out in a 2-dimensional co-ordinate system.

A surface may be defined by the parametric equation

$$\boldsymbol{r} = \boldsymbol{r}(u, v) \tag{2.1}$$

where u and v are arbitrary local co-ordinates. The contour limiting the surface is arbitrary, as sketched in Figure 2.1*a*.

Assume that we have adopted basis functions of the local co-ordinates u and v for the approximation of the surface currents on such a surface. To satisfy the current-continuity equation along the surface-contour line, in the general case both surface-current components must be considered. Therefore it appears to be difficult to construct basis functions which will satisfy the current-continuity equation automatically along the whole length of the surface-contour line. This difficulty can be greatly eased by adopting the surface-contour line which coincides with the local co-ordinate lines. In this case the surface becomes a generalised quadrilateral, defined by

$$u_1 \leqslant u \leqslant u_2 \qquad v_1 \leqslant v \leqslant v_2 \tag{2.2}$$

where u_1, u_2, v_1 and v_2 are arbitrary starting and end co-ordinates of the quadrilateral sides, as shown in Figure 2.1*b*. If we adopt all surface elements to

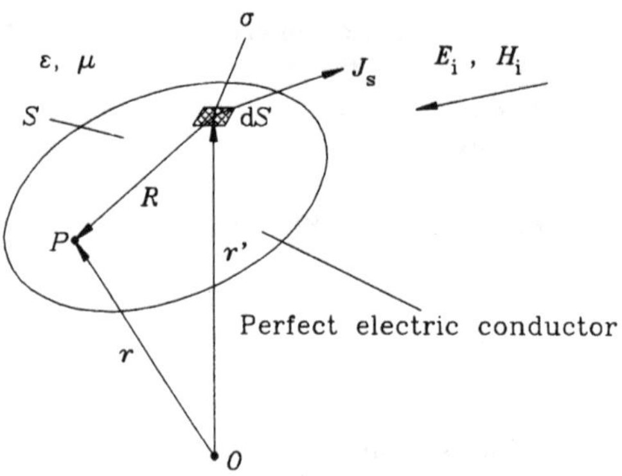

Figure 2.1 Parametric surface element with (a) arbitrary contour line and (b) limits coinciding with local-co-ordinate lines (generalised quadrilateral)

be such generalised quadrilaterals, this greatly simplifies the definition of any surface element. In addition, and perhaps even more important, along any quadrilateral side this also requires only one surface-current component (that which is not tangential to the side considered) in formulating the current-continuity equation.

To illustrate this conclusion, consider an arbitrary n-sided polygon with sides defined with respect to the rectangular $x0y$ co-ordinate system by the equations

$$\boldsymbol{r}\,.\,\boldsymbol{n}_i = c_i \qquad \boldsymbol{r} = x\boldsymbol{i}_x + y\boldsymbol{i}_y \qquad \boldsymbol{n}_i = a_i\boldsymbol{i}_x + b_i\boldsymbol{i}_y \qquad i = 1, 2, \ldots, n \tag{2.3}$$

where $\boldsymbol{n}_i$ is the unit vector normal to the ith polygon side. Let the surface-current-density-vector components over the polygon surface be approximated by the double power series

$$J_{sx} = \sum_{i=1}^{n_x}\sum_{j=1}^{n_y} a_{xij}x^{i-1}y^{j-1} \tag{2.4}$$

$$J_{sy} = \sum_{i=1}^{n_x}\sum_{j=1}^{n_y} a_{yij}x^{i-1}y^{j-1} \tag{2.5}$$

where a_{xij} and a_{yij} are unknown coefficients. The current-continuity equation along any free polygon side (i.e. a side not shared with any other surface element, along which the normal surface-current component must be zero) has the form

$$(J_{sx}\boldsymbol{i}_x + J_{sy}\boldsymbol{i}_y)\,.\,\boldsymbol{n}_i|_{a_ix+b_iy=c_i} = (a_iJ_{sx} + b_iJ_{sy})|_{a_ix+b_iy=c_i} = 0 \tag{2.6}$$

By using eqns. 2.4 and 2.5, the left-hand side of eqn. 2.6 can be written in the form of a polynomial in x or y. The equation thus obtained is equivalent to a system of linear equations in a_{xij} and b_{xij}, the number of which is equal to the degree of the polynomial plus one. In the general case a_i and b_i are not equal to zero, and the degree of the polynomial (the number of equations) is $n_x + n_y - 2$. If a_i or b_i is zero, however, i.e. if the polygon side coincides with the x or y co-ordinate line, the degree of the polynomial is only $n_x - 1$ or $n_y - 1$, respectively. Therefore in such special cases the equations expressing the continuity equation are indeed much simpler than in the general case, and contain only one current component.

2.2.2 Some geometrical quantities important for approximation of surface currents

The u and v co-ordinates are not in general the length co-ordinates. If not, the length elements $\mathrm{d}l_u$ and $\mathrm{d}l_v$ along the co-ordinate lines, corresponding to the increments $\mathrm{d}u$ and $\mathrm{d}v$ of the u and v co-ordinate, respectively, along them, are obtained as (see, for example, Reference 75)

$$\left.\begin{aligned} \mathrm{d}l_u &= e_u\,\mathrm{d}u \\ \mathrm{d}l_v &= e_v\,\mathrm{d}v \end{aligned}\right\} \tag{2.7}$$

where e_u and e_v are the Lamé coefficients:

$$\left.\begin{aligned} e_u &= \left|\frac{\mathrm{d}\boldsymbol{r}}{\mathrm{d}u}\right| \\ e_v &= \left|\frac{\mathrm{d}\boldsymbol{r}}{\mathrm{d}v}\right| \end{aligned}\right\} \tag{2.8}$$

The unit vectors of the local co-ordinate system at any point of the surface element considered are obtained as

$$\left.\begin{aligned} \boldsymbol{i}_u &= \frac{1}{e_u}\frac{\mathrm{d}\boldsymbol{r}}{\mathrm{d}u} \\ \boldsymbol{i}_v &= \frac{1}{e_v}\frac{\mathrm{d}\boldsymbol{r}}{\mathrm{d}v} \end{aligned}\right\} \tag{2.9}$$

Finally, the area element $\mathrm{d}S$ of the surface can be expressed as

$$\mathrm{d}S = \mathrm{d}l_u\,\mathrm{d}l_v \sin\alpha_{uv} = e_u e_v \sin\alpha_{uv}\,\mathrm{d}u\,\mathrm{d}v \tag{2.10}$$

where α_{uv} is the local angle between the u and v co-ordinate lines.

By appropriate choice of the equation (eqn. 2.1) defining the surface element, it is possible not only to obtain a desired flexibility of the elements necessary for efficient approximation of the surface, but also to enable a simpler evaluation of the quantities in eqns. 2.7–2.10 describing the local co-ordinate system. Some Sections that follow will be devoted to precisely that problem.

2.3 Degenerate forms of generalised quadrilaterals

2.3.1 Generalised wires

Wire structures assembled from arbitrarily interconnected wire segments represent a special class of metallic antennas and scatterers. In the general case wire segments are curvilinear, and may be solid or hollow, of constant or

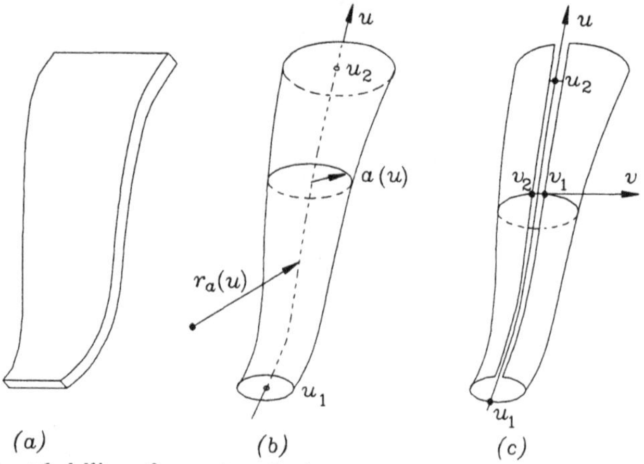

Figure 2.2 *Modelling of geometry of wires*

a Wire of cross-section in the form of thin rectangle
b Approximately equivalent wire of circular cross-section (generalised wire)
c Generalised wire represented as generalised quadrilateral

variable cross-section (e.g. of the forms shown in Figure 2.2). In the electromagnetic analysis, wire segments can usually be considered to be made of a perfect conductor.

A wire is said to be thin if the greatest dimension of its cross-section is much smaller than the wavelength of the incident field, the length of the segment and the radius of curvature of the wire axis. The distribution of the total current along such a wire does not change significantly if an equivalent circular cross-section is used instead of the original one. The equivalent radius is determined from the requirement that the quasistatic energy in the vicinity of the wire, per unit wire length, should be equal in the original and equivalent cases (see, for example, Reference 34). For example, for an infinitely thin strip of width w the equivalent radius is known to be $a = w/4$.

The surface-current density along thin wires of circular cross-section is approximately the same around the circumference of any cross-section of the wire, and has no circumferential component. The same is true for symmetrically excited bodies of revolution, so that the analysis of such systems is essentially the same as that of thin wires. The same analysis can also be applied to regions of axially symmetrical abrupt changes in the wire diameter and of the wire ends. Finally, a frill with radial electric surface currents can also be considered as a degenerate form of such a wire, and can be used for the treatment of wire-to-plate junctions. Although the degenerate forms just mentioned are not wirelike at all, we shall term them briefly 'generalised wires' [64–66].

The geometry of a generalised wire segment is completely determined by parametric equations of its axis and its radius:

$$\mathbf{r}_a = \mathbf{r}_a(u) \qquad a = a(u) \qquad u_1 \leqslant u \leqslant u_2 \tag{2.11}$$

(see Figure 2.2*b*). Note that such a wire segment can also be visualised as a degenerate form of a generalised quadrilateral, as illustrated in Figure 2.2*c*.

Therefore any theory including the electromagnetic analysis of generalised quadrilaterals can be applied to generalised wires. However, the current distribution along generalised wires is 1-dimensional (one current component depending on one co-ordinate only), and the current distribution over generalised quadrilaterals is, generally speaking, 2-dimensional (two current components depending on two co-ordinates). This fact can be used to make the analysis of generalised wires formally the same, but significantly simpler and more rapid than that of generalised quadrilaterals. Therefore in what follows we shall also consider specific methods for the analysis of generalised wires.

2.3.2 *Generalised triangles*

If two adjacent nodes of a generalised quadrilateral coincide (e.g. node $u = u_1$, $v = v_1$ and node $u = u_1$, $v = v_2$, as shown in Figure 2.3), it degenerates into a curved curvilinear triangle, i.e. a 'generalised triangle' [67]. Such triangles can alternatively be used as elements for the approximation of arbitrary geometical surfaces.

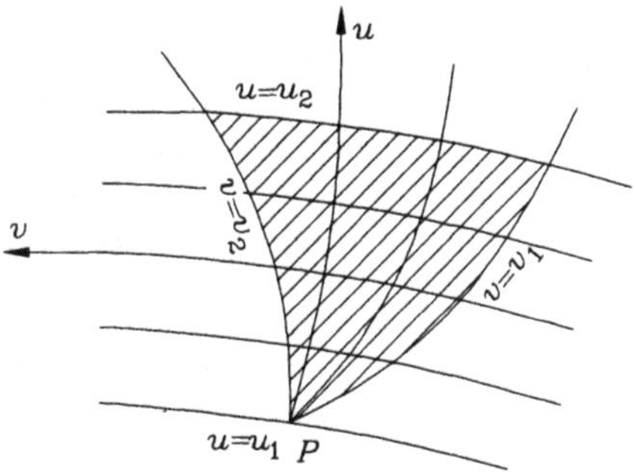

Figure 2.3 Generalised triangle

It is a simple matter to understand that the surface-current density over generalised triangles cannot be represented at all points in terms of its u and v components alone, as can be done in the case of generalised quadrilaterals. The singular point is the point P in Figure 2.3. At that point the u-co-ordinate lines intersect, so that there the unit vector $\boldsymbol{i}_u$ is not uniquely defined. Therefore the u component of the surface-current-density vector at such a point can only be zero. However, the v component of the surface-current-density vector cannot represent the total current, except in special cases. A remedy is to represent the surface-current-density vector over triangles by means of three components. This will be discussed in more detail in Chapter 3.

It follows that at least part of the theory valid for generalised quadrilaterals is not applicable to generalised triangles. The necessary modifications of the general theory to generalised triangles will therefore also be discussed.

2.3.3 *Bodies of revolution and bodies of translation*

Bodies of revolution and bodies of translation (acronyms BOR and BOT, respectively, are sometimes used for such bodies) represent special classes of

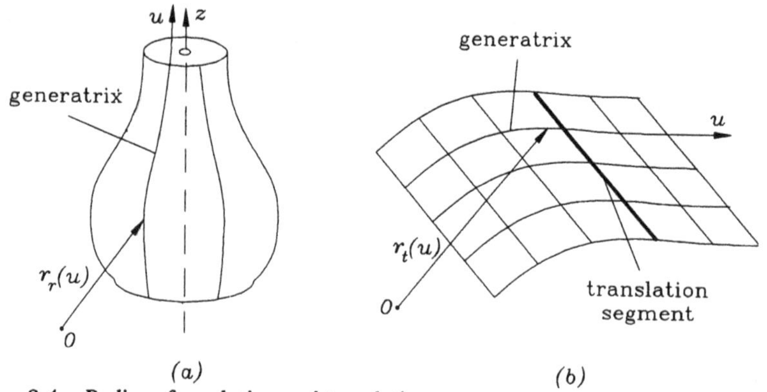

Figure 2.4 Bodies of revolution and translation

a Body of revolution
b Body of translation

metallic antennas and scatterers. They have been analysed quite frequently (see, for example, References 55, 76 and 77), for the following reasons. First, many structures of interest can be modelled as bodies of revolution (e.g. rockets, satellites etc.) or of translation (e.g. rectangular plates, noncircular cylinders etc.). Secondly, the geometry of these two classes of bodies can be described in a very simple manner. [The geometry of a body of revolution is completely determined by the parametric equation of its generatrix $\mathbf{r}_r(u)$ and the axis of revolution, as shown in Figure 2.4*a*. The geometry of a body of translation is completely determined by the parametric equation of its generatrix $\mathbf{r}_t(u)$, and the position and length of the segment that makes the body by translation, as shown in Figure 2.4*b*.] Thirdly, and most importantly, the two classes of body appeared simpler to analyse than the general structure. However, with the method of analysis of general structures presented in this book such bodies are only special cases of these general structures. Therefore bodies of revolution and bodies of translation will not be considered separately.

2.4 Exact modelling of geometry by generalised quadrilaterals

Exact modelling of the geometry of a structure by generalised quadrilaterals is possible only if the structure surface can be described exactly by one or more equations. If these equations, at the same time, represent the parametric equations of generalised quadrilaterals, the modelling does not need any further procedure. If they do not, which is usually the case, it is necessary to obtain the parametric equations of generalised quadrilaterals by means of known exact equations of the structure surface.

As an example, let us attempt to define generalised quadrilaterals that follow a spherical shape exactly. This can be done in many ways. Let us adopt the following procedure [67]. Let, first, a cube be inscribed into the sphere. For all the six cube sides the local (u, v) co-ordinate systems are next defined, with

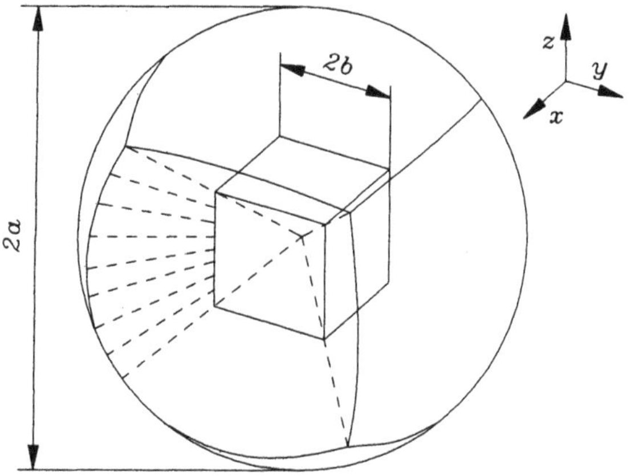

Figure 2.5 Exact modelling of a sphere by generalised quadrilaterals

$-1 \leqslant u,\ v \leqslant 1$. Finally, the cube sides together with the co-ordinate systems are projected onto the sphere surface, with the common sphere and the cube centre as the projection centre.

To illustrate this procedure, consider the upper side of the cube sketched in Figure 2.5. Let the radius of the sphere be a, and the side length of the cube be $2b$. The parametric equation of the cube side considered can be written as

$$\boldsymbol{r}_b(u, v) = \boldsymbol{i}_z b + \boldsymbol{i}_x bu + \boldsymbol{i}_y bv \qquad (2.12)$$

The parametric equation of the corresponding generalised quadrilateral obtained by projecting this cube side onto the sphere surface is now obtained as

$$\boldsymbol{r}_a(u, v) = a\,\frac{\boldsymbol{r}_b(u, v)}{|\boldsymbol{r}_b(u, v)|} = a\,\frac{\boldsymbol{i}_z + \boldsymbol{i}_x u + \boldsymbol{i}_y v}{\sqrt{(1 + u^2 + v^2)}} \qquad (2.13)$$

Using this expression, it is not difficult to compute other geometrical quantities of interest, e.g. the Lamé coefficients etc.

Finally, note that if instead of the cube we begin with a tetrahedron inscribed in the sphere, by following a similar procedure the sphere can be modelled exactly by generalised triangles.

2.5 Notes on approximate modelling of geometry

2.5.1 Introduction

Approximate modelling of the geometry of electromagnetic structures is not as simple as it might look. At first glance, the only essential requirement for the modelling appears to be that the electromagnetic properties of the approximate structure be sufficiently close to those of the original. Obviously, the exact meaning of 'sufficiently close' will differ from one problem to another, and must usually be adopted on the basis of engineering judgment. In addition, different electromagnetic quantities require different degrees of approximation of geometry to achieve acceptable values of the quantities. For example, the near field requires more accurate modelling of geometry and current distribution for the same accuracy than the far field. More precisely, accurate evaluation of the far field is possible also if the approximate surface has edges that do not exist in the actual surface; i.e. only continuity of the surface, and not the continuity of the first derivative of the surface, suffices. Accurate evaluation of the near field, however, also requires continuity of the first derivative of the surface. Therefore the required accuracy of modelling of geometry depends also on the electromagnetic quantity we are specifically interested in.

Even if we specify the maximum permitted difference in performance between the actual structure and its geometrical approximation, there are still many aspects of the problem to be considered. To understand this, assume that we have adopted an approximate geometry of the structure. The next step, to be considered in detail in the following Chapters, is to determine the currents in the structure. Except in a few cases, they can be determined only approximately. Therefore we almost always have the following situation: we approximate the structure geometry, and then we search for approximate currents in the approximate model of the structure. Obviously, many questions may be asked concerning the final accuracy or acceptability of the results obtained by this sequence of approximations. The purpose of this Section is to discuss briefly some of the problems that the authors found most important, but also to give certain guidelines relating to the required accuracy of modelling of the geometry of electromagnetic stuctures.

Generalised quadrilaterals and generalised wires can approximate many structures quite efficiently, but the approximate structure most frequently contains edges (at interconnections of generalised quadrilaterals) and/or broken cones (at interconnections of generalised wîres), while the original structure usually does not. Theoretically, the charge and current density at such edges is infinite. Fortunately, in most instances these and similar problems are of almost no practical importance, since the current distribution can be determined only approximately. Therefore, for example, an edge in the model combined with the appproximate current distribution does not result in a singular charge density or electric-field strength, except if the type of current approximation is intentionally adopted in a form which enables such singularities to exist in the solution. Therefore the edges between two generalised quadrilaterals and abrupt changes in the direction of wire segments connected in a row in the approximate structure may have only a minor effect on the final approximate solution, provided that the order of approximation for the currents is not excessive.

As the next problem, note that it is not possible in all cases to approximate a continuous surface by means of generalised quadrilaterals and generalised wires. For example, the interconnection of two generalised wires is not a continuous surface (unless they are coaxial and of the same radius at the interconnection), and neither is the interconnection of a generalised wire and a generalised quadrilateral (except in very special cases). Therefore, the approximate metallic structure will usually be discontinuous at some points. Fortunately, this is again a problem only formally, provided that the region of such imperfections is electrically small. Specifically, if we stipulate the current-continuity equation at such places (which is usually done in most methods of solution), this discontinuity is not 'seen' in the final solution for the currents. This is a very important conclusion, as it permits significant simplifications in approximate modelling of metallic structures.

2.5.2 Description of difference between approximate and actual structures

First, it is necessary to define a quantity which describes the difference between the actual and approximate structures. There are many ways of doing this. For example, this quantity can describe the difference integrally, locally, as an average (arithmetic or geometric) of a number of samples etc. Whatever the definition, however, it is necessary to adopt a measure of the difference between the actual and approximate structures at any point of the actual structure. For a smooth original surface or a wire without sharp bends, we define this to be the distance between corresponding points of the two, perpendicular to the original structure surface, namely to the original wire, and term this quantity briefly the 'deviation' (of the approximate structure with respect to the original). For example, if we approximate a circular loop by a square loop circumscribed about the original, the deviation is zero at the four common points, and the largest deviation is at the four points that are at 45° from the first set of points, equal to $(\sqrt{2}-1)$ times the radius of the loop. As another example, if we approximate a sphere by a cube circumscribed about the sphere, the largest deviation is equal to $(\sqrt{3}-1)$ times the sphere radius.

Obviously, the largest deviation is not sufficient in itself to describe the overall difference in the two structures. Nevertheless, we shall use it in what follows as an indication of this difference. Note that in most cases this will be a very conservative indicator.

2.5.3 The basic classes of shapes for geometrical modelling of metallic antennas and scatterers

Our aim is to develop an efficient method for the electromagnetic analysis of arbitrary metallic structures. The modelling of geometry can, of course, be perfomed by small patches, irrespective of the shape of the structure. This is not practical because it usually requires very large numbers of patches (e.g. in approximating the surface of a wire) and, consequently, very large number of unknowns when solving for current distribution. The other extreme is to consider a wide variety of shapes which can serve as building blocks of a large class of body shapes. This is not practical because it is neither simple to define such shapes, nor easy to determine currents over them.

A compromise which seems reasonable is to define the smallest possible number of basic shapes that enable efficient approximation of as a wide class of practical structures as possible. In the opinion of the authors, most of the structures which are of interest can be approximated quite efficiently by only two types of basic shape: generalised quadrilaterals (intended for the approximation of surfaces) and generalised wires (intended for the approximation of wires and related discontinuities), with their degenerate forms. It will be shown in Chapter 3 that it is relatively simple to determine current distribution over these two basic shapes, so that a solution for the currents in a structure assembled from them is also relatively simple. Therefore this book will be based on the approximation of any metallic structure by these two basic shapes only. Note that it is quite simple to represent a metallic surface or a wire defined by a set of points by means of these two basic shapes.

Only rarely it is possible to define a structure geometry by analytical formulae. Instead, this is usually done by tabulated data, describing the structure by the position vectors of a number of the structure points (and possibly the tangent unit vectors and the radii of curvature at these points etc.). It is therefore necessary that the approximation of the structure geometry be such that the representation of the actual structure by tabulated data can easily and efficiently be utilised.

2.5.4 Surface-patch, wire-segment and wire-grid modelling

There are several approaches to the approximation of geometry of combined-wire-and-plate structures. In the general case, any surface (wire or plate surface) can be modelled by planar or curved surface patches. This kind of modelling is known as 'surface-patch modelling'.

However, although applicable for modelling of arbitary surfaces, surface-patch modelling is obviously impractical for modelling of thin wires. Instead, wires (including curved wires, wire ends and abrupt changes of wire radius) can be modelled much more efficiently by a sequence of appropriate generalised-wire segments. This type of modelling we shall term 'wire-segment modelling'.

Since it is possible to approximate any metallic surface by a wire net or grid, wire-segment modelling can also be applied to the approximation of surfaces. In that case it is referred to as 'wire-grid modelling'. However, although simple in principle, wire-grid modelling of surfaces has two major deficiencies. On the one hand, there is no rule as to how dense the grid need be to obtain accurate results for the current distribution. On the other hand, the number of unknowns in wire-grid modelling tends to be large for even electrically small metallic bodies.

On the basis of this short comparison of the possible types of modelling geometry, surface-patch modelling obviously appears to be optimal for modelling of plates, and wire-segment modelling for modelling of wirelike structures.

2.6 Approximate modelling of generalised wires

2.6.1 Approximation of wires by spline curves

Approximation of generalised wires consists in the approximation of the parametric equations of its radius and its axis. Note that the first equation is a scalar function of one unknown (the co-ordinate along the wire axis), and the second is a vector function of the same unknown.

Consider first the approximation of the parametric equation of the wire radius. This equation can be approximated by a polynomial in the standard manner:

$$a(u) = \sum_{i=0}^{n} p_i u^i \qquad (2.14)$$

where n is the order of approximation, and the scalar coefficients p_i can be determined in different ways starting, possibly, also from tabulated data. The simplest way to determine these coefficients is to equate this polynomial with known values of the radius at $(n+1)$ points, $a_i = a(u_i)$, $i = 1, \ldots, (n+1)$, and then to solve the system of linear equations thus obtained for the unknown coefficients p_i. In this particular case the polynomial in eqn. 2.14 can be written in the form

$$a(u) = \sum_{i=1}^{n+1} q_i(u) a_i \tag{2.15}$$

where

$$q_i(u) = \frac{(u-u_1)\ldots(u-u_{i-1})(u-u_{i+1})\ldots(u-u_{n+1})}{(u_i-u_1)\ldots(u_i-u_{i-1})(u_i-u_{i+1})\ldots(u_i-u_{n+1})} \tag{2.16}$$

is known as the Lagrange interpolation polynomial (e.g. Reference 78).

In a similar way the parametric equation of the wire axis can be approximated by a vector polynomial:

$$\boldsymbol{r}_a(u) = \sum_{i=0}^{n} \boldsymbol{P}_i u^i \tag{2.17}$$

where $\boldsymbol{P}_i$ are vector coefficients that can also be determined in different ways, possibly starting from tabulated data. The simplest way of determining these coefficients is again to equate the vector polynomial with the known values of the position vectors of the wire axis at $(n+1)$ points, $\boldsymbol{r}_{ai} = \boldsymbol{r}_a(u_i)$, $i = 1, \ldots, (n+1)$, and to solve the system of linear equations thus obtained for the unknown coefficients $\boldsymbol{P}_i$. In that special case the polynomial in eqn. 2.17 can be written in the form

$$\boldsymbol{r}_a(u) = \sum_{i=1}^{n+1} q_i(u) \boldsymbol{r}_{ai} \tag{2.18}$$

where the functions $q_i(u)$ are given in eqn. 2.16.

Many other interpolation polynomials are available. Some take into account not only tabulated values of the function being approximated, but also tabulated values of the first- and higher-order derivatives of this function. Others are characterised by specific distributions of points used for the interpolation. In all instances, the goal is to obtain the best approximation (in some sense) of the original curve. However, all interpolation polynomials show undesirable oscillations if higher-order approximations are used. To avoid this, the segments are usually subdivided into a set of shorter subsegments, each described by a separate low-order ($n \leqslant 3$) interpolation polynomial called a 'spline'. The curves obtained in this manner are known as 'spline curves'.

The expressions for some specific splines, such as those in eqns. 2.15, 2.16 and 2.18, can easily be transformed into a simpler common form given by eqns. 2.14 and 2.17. It is convenient to begin modelling of the geometry of wires with the expressions for splines, and to end with the common polynomial expressions that, in turn, can in a later step be used for efficient approximation of current distribution. Therefore the procedure for determining the currents depends only on the order of the applied spline, and not on its type (because all are transfomed into simple polynomials prior to determining current distribution). With this in mind and taking into account the notes in Section 2.5, certain recommendations concerning the order and type of the spline can be suggested.

The Lagrange spline, with the first tabulated point at the beginning and the last at the end of the interpolation segment, can provide continuity of the wire local axis and of its radius, which is satisfactory for most purposes. Better modelling of geometry can be obtained by using other splines, such as B-spline or beta-spline, used frequently in computer graphics [79]. These splines can not only provide continuity of the wire axis and its radius, but also eliminate the possibility of a broken (although continuous) line approximating the wire axis. However, this improvement in modelling of geometry is usually unnecessary in the determination of current distribution along generalised metallic wire antennas and scatterers. In addition, evaluation of these splines tends to be much more complicated than evaluation of the Lagrange splines. Therefore simple Lagrange splines appear to be the most convenient basic splines for modelling of geometry in the electromagnetic analysis of generalised-wire antennas and scatterers.

Possible degrees of the Lagrange splines used for modelling of geometry of generalised wires are one (linear spline), two (quadratic spline) and three (cubic spline). Linear splines correspond to right truncated cones, a structure very convenient for modelling of geometry of generalised wires. This is considered in more detail in Section 2.6.2.

2.6.2 Right truncated cones

Wire structures are usually assembled from straight segments of constant (possibly equivalent) radius. Therefore only this type of wire structure has been analysed by most researchers of wire antennas and scatterers. This model can also be used to approximate curved wires, as a sequence of short straight segments. However, it cannot be used for accurate approximation of wires of continuously variable cross-section, except in the form sketched in Figure 2.6*a*,

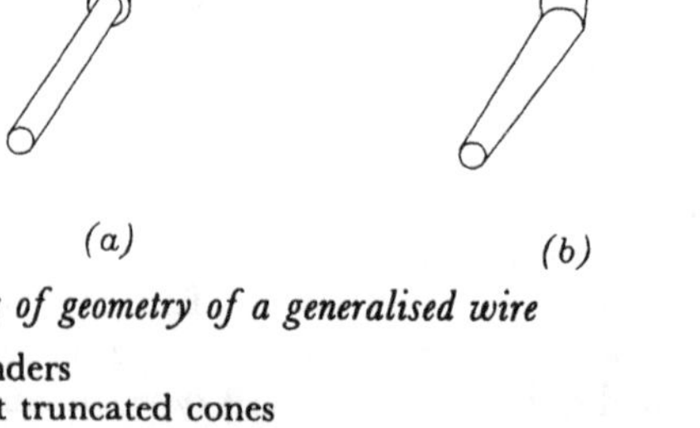

(a) (b)

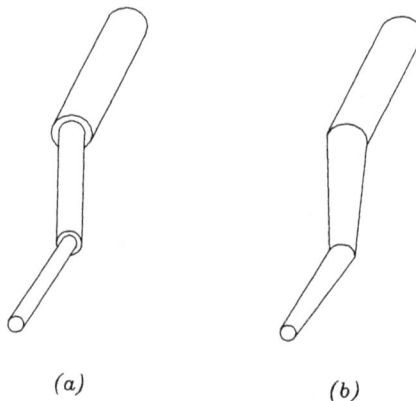

Figure 2.6 Modelling of geometry of a generalised wire

a By cylinders
b By right truncated cones

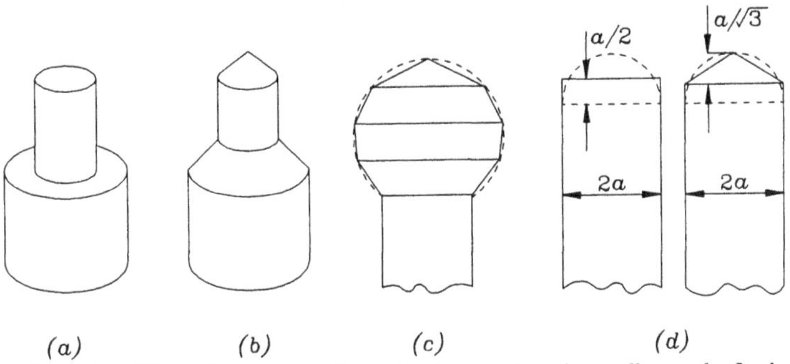

Figure 2.7 *Modelling of geometry of abrupt changes in wire radius and of wire ends*

a A flat junction and end
b A conical junction and end
c Approximation of a spherical end by a sequence of short conical segments
d Flat and conical end equivalent to a hemispherical end

which represents piecewise-cylindrical approximation of the wire shown in Figure 2.2*b*. This difficulty can be partly avoided if right truncated cones are used instead of cylinders [64–66], as indicated in Figure 2.6*b*.

In particular cases, a truncated cone degenerates into a right cylinder, an ordinary cone, a flat disc and a frill. These structures can be used for modelling cylindrical wires with flat (frill-like) or conical changes of the wire radius, as well as of flat and conical wire ends, as indicated in Figures 2.7*a* and 2.7*b*. A frill can also be used for modelling of wire-to-plate junctions, as will be shown in Section 2.8.1. It is a simple matter to understand that any axially symmetrical segment of a generalised wire can be either exactly or approximately modelled by right truncated cones. An example is shown in Figure 2.7*c*, where a spherical wire end is approximated by a sequence of short conical segments. Note, however, that in most instances such an accurate representation of the wire end is not necessary. Wire ends of simpler shapes can be substituted by approximately equivalent flat or conical ends [11, 80], as shown in Figure 2.7*d*. The equivalent end is obtained by slightly changing the length of the wire and replacing the original wire end by a flat or conical one. The size of the equivalent end is obtained by requesting that the surfaces of the original and equivalent ends have equal area. In that case, approximately the same amount of charge is located on both, so that in both cases the field along the wire due to these charges is also approximately the same. In a similar way, it is possible to approximate equivalent flat and conical abrupt changes in wire radius.

It will be seen that, except for insignificant additional analytical effort, the determination of current distribution along truncated cones does not require longer computing time than that along cylindrical wires. However, the approximation of generalised wires by higher-order (quadratic and cubic) splines increases considerably both the analytical effort and the computing time. Therefore right truncated cones seem to be the optimal basic elements for modelling of geometry and the analysis of structures assembled from generalised wires.

A right truncated cone is determined by the position vectors and the radii of

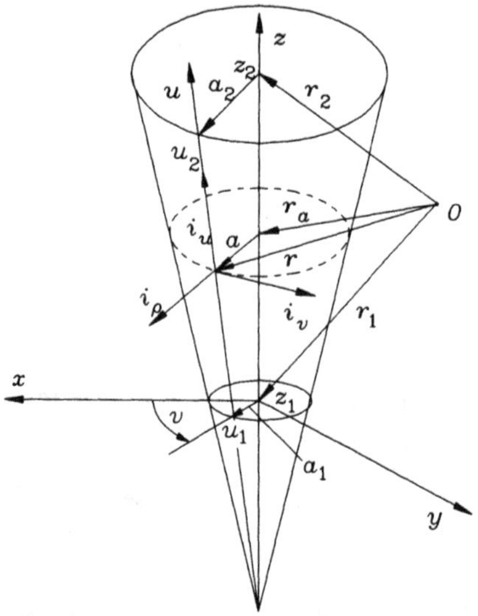

Figure 2.8 Elements for description of geometry of a right truncated cone

its beginning and its end, $\boldsymbol{r}_1$ and a_1, and $\boldsymbol{r}_2$ and a_2, respectively, as shown in Figure 2.8.

The parametric equation of its surface can be written in the form

$$\boldsymbol{r}(u, v) = \boldsymbol{r}_a(u) + a(u)\boldsymbol{i}_\rho(v) \qquad u_1 \leqslant u \leqslant u_2 \qquad v_1 \leqslant v \leqslant v_2 \tag{2.19}$$

where $\boldsymbol{r}_a(u)$ and $a(u)$ are the parametric equations of the cone axis and its local radius,

$$\boldsymbol{r}_a(u) = \boldsymbol{r}_1 + (u - u_1)\frac{\boldsymbol{r}_2 - \boldsymbol{r}_1}{u_2 - u_1} \qquad u_1 \leqslant u \leqslant u_2 \tag{2.20}$$

$$a(u) = a_1 + (u - u_1)\frac{a_2 - a_1}{u_2 - u_1} \qquad u_1 \leqslant u \leqslant u_2 \tag{2.21}$$

In eqns. 2.19–2.21, u is the local co-ordinate along the reference-cone generatrix, v is the local co-ordinate about the cone axis, measured from the x axis, and $\boldsymbol{i}_\rho(v)$ is the radial unit vector in the local-co-ordinate system, perpendicular to the cone axis, as shown in Figure 2.8. [The reason for adopting the u co-ordinate rather than the z co-ordinate (that is along the cone axis) is as follows. It will be assumed that the surface-current-density vector has only the u component, which cannot be expressed in terms of the z co-ordinate if the u and z axes are normal to one another.] In the general case u and v are not length co-ordinates. The co-ordinates u_1 and u_2 can be adopted arbitrarily, as can the unit vector $\boldsymbol{i}_\rho(0) = \boldsymbol{i}_x$, which represents the reference axis for the v co-ordinate.

Starting from the expressions in eqns. 2.19–2.21, it is not difficult to obtain any other quantity of interest in this case. As an example, consider again the Lamé coefficients:

$$\left.\begin{aligned} e_u &= \left|\frac{d\boldsymbol{r}}{du}\right| \\ \frac{d\boldsymbol{r}}{du} &= \frac{\boldsymbol{r}_2 - \boldsymbol{r}_1}{u_2 - u_1} + \frac{a_2 - a_1}{u_2 - u_1}\boldsymbol{i}_\rho(v) \end{aligned}\right\} \tag{2.22}$$

$$\left.\begin{aligned} e_v &= \left|\frac{\mathrm{d}\boldsymbol{r}}{\mathrm{d}v}\right| = a(u) \\ \frac{\mathrm{d}\boldsymbol{r}}{\mathrm{d}v} &= a(u)\,\boldsymbol{i}_\phi(v) \end{aligned}\right\} \tag{2.23}$$

To simplify the analysis, it is convenient that the starting and end values of the u co-ordinate of a truncated cone be adopted as -1 and $+1$, respectively.

2.6.3 *Piecewise-cylindrical approximation of wires*

Segmentation of wires is needed in two cases. First, it is needed for electrically long wires. The number of functions used for current approximation along a wire is roughly proportional to its electrical length. Therefore, if low-order expansions are used for the appproximation of current distribution along a wire segment, the electrical length of the segment is limited. Although there are notable exceptions (e.g. trigonometric plus polynomial current approximation for current along straight-wire segments [81, 33, 66]), entire-domain current approximations cannot usually be used for segments much longer than one or two wavelengths. Subdomain basis functions cannot be successfully applied for segments longer than 0.1–0.25λ [28]. Therefore even cylindrical wire segments, if too long, must be subdivided into a number of subsegments.

Segmentation of wires is also needed if they are curved. Satisfactory results for the antenna impedance, current distribution etc. can be obtained only if the deviation (as defined in Section 2.5.2) of the segmented wire axis and of its radius is small enough. The definition of 'small enough' is different for electrically short and electrically long wires. For electrically short wires, deviation of their axis should be small when compared with the length of the wire (or some other characteristic length), e.g. less than about 10% of the wire length. For electrically longer wires, deviation of their axis should be small in comparison with the wavelength, for example it should not exceed 1/32nd or 1/16th of the wavelength. Of course, in both cases the relative deviation of the radius should not be greater than the relative deviation of the axis. The above recommendations are fairly obvious: for electrically short wires there is no sense in using the wavelength as a reference, while for electrically longer wires this must be done.

Having in mind the maximum electrical length of a truncated cone and the maximum deviations allowed, many segmentation procedures can be constructed. A good procedure is one which permits relatively uniform division of the wire with a small number of segments. Unfortunately, although segmentation of wires may look easy, there is no simple algorithm that could provide the optimum segmentation in all cases. Experience therefore plays an important role in obtaining the optimum piecewise-cylindrical approximation of curved wires.

For example, assume that the chosen segmentation procedure starts by adopting one end of the wire as the beginning of the first truncated cone. Let the end of this cone be on the curved wire axis, its length determined according to the above criteria for electrically long curved wires. We choose the beginning of the second cone to be at the end of the first etc. Such a procedure is satisfactory in most cases. However, note that, with this procedure for approximation of curved wires, each truncated cone is shorter than the length of the corresponding curved wire

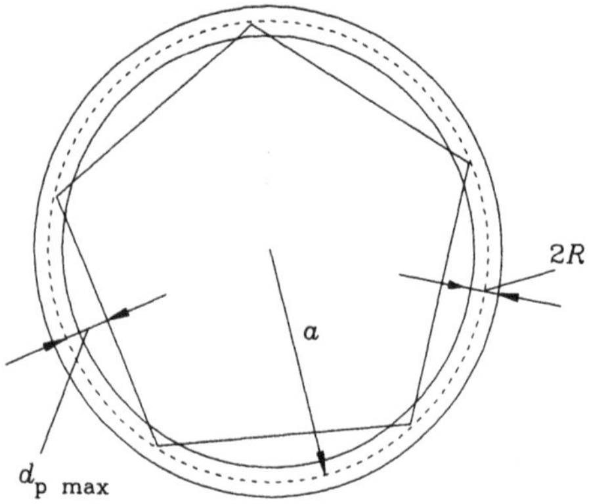

Figure 2.9 Approximation of a ring scatterer by an inscribed polygon

segment, so that the total length of the piecewise-cylindrical approximation obtained is shorter than the original curved wire. This difference in length can introduce sizable errors in the results of the analysis if the wire is of approximately resonant length. This can be avoided if all cylindrical segments are chosen to be the same length as the corresponding curved wire segment (of course, in that case the cylindrical segment ends cannot all be on the curved wire axis).

As a specific example, consider a circular ring scatterer of radius a, made of thin wire, shown in Figure 2.9. Let the ring be approximated by an n-sided polygon inscribed in the ring. Table 2.1 gives the following quantities for $n = 3, 4, \ldots, 10$:

(*a*) the maximum deviation of the polygon with respect to the ring, normalised with respect to the ring radius, $d_{p\,\max}/a$
(*b*) the maximum electrical radius of the ring, $a_{\max}/\lambda$, for which $d_{p\,\max}/a$ is 1/24th of the wavelength
(*c*) the electrical length of one polygon side corresponding to the maximum electrical radius of the ring, $l_{\max}/\lambda$
(*d*) the corresponding electrical difference in the circumferences, $d_{c\,\max}/\lambda = (2\pi a_{\max} - nl_{\max})/\lambda$.

It can be seen from Table 2.1 that, for rings of electrically small and medium radii ($a \leqslant 0.3\lambda$), a reasonable approximation can be obtained with $n = 6$ segments. For electrically larger radii, it is necessary to use more than six segments. Note that $a_{\max}/\lambda$ is not proportional to n. It can easily be shown that for larger n (e.g. $n > 6$) this ratio is approximately proportional to n^2, i.e. as the electrical radius of the ring is increased the number of segments should increase approximately as the square root of the radius. This is why the maximum electrical length of the polygon side (in the above sense) increases for larger electrical radii of the ring. Its value indicates that subdomain basis functions cannot be used for analysis of larger rings modelled by the minimum number of segments, i.e. additional segmentation is needed. Finally, note that the difference in circumference is as large as about $\lambda/10$, which can cause additional errors. It is obvious that modelling of the geometry of the ring can be improved if a polygon of the same circumference is used instead of the inscribed polygon.

Table 2.1 Data illustrating approximation of thin-wire circular ring scatterer of radius a by n-sided polygon inscribed in the ring

n	$d_{p\,max}/a$	a_{max}/λ	l_{max}/λ	$d_{c\,max}/\lambda$
3	0.500	0.083	0.144	0.091
4	0.293	0.142	0.201	0.089
5	0.191	0.218	0.256	0.088
6	0.134	0.311	0.311	0.088
7	0.099	0.421	0.365	0.088
8	0.076	0.547	0.419	0.088
9	0.060	0.691	0.473	0.088
10	0.049	0.851	0.536	0.088

$d_{p\,max}/a$ = maximum deviation of the polygon with respect to the ring, normalised with respect to the ring radius

a_{max}/λ = maximum electrical radius of the ring for which $d_{p\,max}/a = 1/24$th of the wavelength

l_{max}/λ = electrical length of one polygon side corresponding to the maximum electrical radius of the ring

$d_{c\,max}/\lambda = (2\pi a_{max} - n l_{max})/\lambda$, the corresponding electrical difference in the circumferences

2.7 Approximate modelling of generalised quadrilaterals

2.7.1 Approximation of surfaces by spline quadrilaterals

In analogy with approximate modelling of generalised wires, the parametric equation of a generalised quadrilateral can be approximated by a vector polynomial as

$$\boldsymbol{r}(u, v) = \sum_{i=0}^{n_u} \sum_{j=0}^{n_v} \boldsymbol{P}_{ij} u^i v^j \tag{2.24}$$

The $\boldsymbol{P}_{ij}$ are vector coefficients that can be determined in different ways if one starts from tabulated data. The simplest way is to equate the vector polynomial with known values of the position vectors at $(n_u + 1) \times (n_v + 1)$ points over the quadrilateral surface, $\boldsymbol{r}_{ij} = \boldsymbol{r}(u_i, v_j)$, $i = 0, 1, \ldots, n_u$, $j = 0, 1, \ldots, n_v$, and to solve the system of linear equations thus obtained. In that special case the polynomial in eqn. 2.24 can be written in the form

$$\boldsymbol{r}(u, v) = \sum_{i=1}^{n_u+1} \sum_{j=1}^{n_v+1} q_i(u) q_j(v)\, \boldsymbol{r}_{ij} \tag{2.25}$$

where q_i and q_j are given by eqn. 2.16.

In analogy with the choice of spline for modelling of generalised wires, the

2-dimensional spline of the Lagrange type appears to be the optimal choice as the basic spline for modelling of geometry of generalised quadrilaterals. Possible orders of this 2-dimensional spline are one (bilinear spline), two (biquadratic spline) and three (bicubic spline). Among these, a bilinear spline surface appears to be particularly convenient for the approximation of geometry of generalised quadrilaterals. This is explained in detail in Section 2.7.2.

2.7.2 Bilinear surfaces

Apart from wires, antennas and scatterers are most often assembled from flat, or approximately flat, metallic plates. Therefore approximation of arbitrary surfaces by flat quadrilaterals and triangles was almost the only approach in modelling the geometry of metallic antennas and scatterers without wires [46, 49]. Such surface elements permit quite simple descriptions of the structure being approximated if flat, or practically flat, plates are considered. Flat quadrilateral surface elements are not suitable for modelling curved surfaces, because they cannot be defined by four arbitrary points in space, unless the four points are in one plane, while triangles can also be used for modelling curved surfaces. However, as explained in Section 2.3.2, three current components need to be specified over triangular surface elements. The difficulties relating to both flat quadrilateral and triangular surface elements can be avoided if bilinear surfaces are used instead of flat quadrilaterals or triangles [64, 65, 67]. Note that this type of surface element can be made to degenerate into a flat quadrilateral or a flat triangle.

It will be seen that, with relatively little additional analytical effort, determination of current distributions over bilinear surfaces does not require longer computing time than that over flat rectangular or triangular surfaces. However, the application of higher-order (biquadratic and bicubic) spline quadrilaterals greatly increases both the analytical effort and the computing time. Therefore bilinear surface elements seem to be optimal in many applications as basic elements for modelling of geometry encountered in the analysis of a wide class of structures assembled from generalised quadrilaterals.

A bilinear surface is, in general, a nonplanar curvilinear quadrilateral, which is defined uniquely by its four vertices. The parametric equation of such a quadrilateral can be written in the form

$$\boldsymbol{r}(u, v) = \frac{1}{\Delta u \Delta v}\{\boldsymbol{r}_{11}(u_2-u)(v_2-v)+\boldsymbol{r}_{12}(u_2-u)(v-v_1) + \boldsymbol{r}_{21}(u-u_1)(v_2-v)+\boldsymbol{r}_{22}(u-u_1)(v-v_1)\} \tag{2.26}$$

$$\Delta u = u_2 - u_1 \qquad \Delta v = v_2 - v_1 \tag{2.27}$$

where $\boldsymbol{r}_{11}$, $\boldsymbol{r}_{12}$, $\boldsymbol{r}_{21}$ and $\boldsymbol{r}_{22}$ are the position vectors of its vertices, and u_1, u_2, v_1 and v_2 are arbitrary starting and end co-ordinates of the quadrilateral sides in the local-co-ordinate system adopted, as sketched in Figure 2.10.

After elementary transformations this equation can be written as

$$\boldsymbol{r}(u, v) = \boldsymbol{r}_c + \boldsymbol{r}_u u + \boldsymbol{r}_v v + \boldsymbol{r}_{uv} uv \qquad u_1 \leqslant u \leqslant u_2 \qquad v_1 \leqslant v \leqslant v_2 \tag{2.28}$$

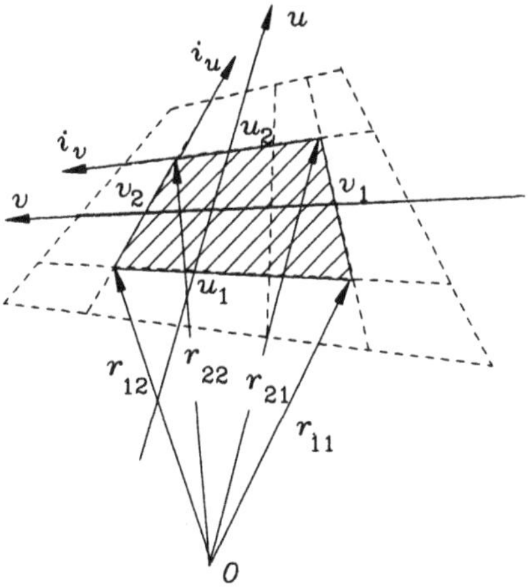

Figure 2.10 Elements necessary for description of geometry of a bilinear surface

$\boldsymbol{r}_c$, $\boldsymbol{r}_u$, $\boldsymbol{r}_v$ and $\boldsymbol{r}_{uv}$ being constant vectors, given by

$$\boldsymbol{r}_c = \frac{1}{\Delta u \Delta v}\left(\boldsymbol{r}_{11}u_2v_2 - \boldsymbol{r}_{12}u_2v_1 - \boldsymbol{r}_{21}u_1v_2 + \boldsymbol{r}_{22}u_1v_1\right) \tag{2.29}$$

$$\boldsymbol{r}_u = \frac{1}{\Delta u \Delta v}\left(-\boldsymbol{r}_{11}v_2 + \boldsymbol{r}_{12}v_1 + \boldsymbol{r}_{21}v_2 - \boldsymbol{r}_{22}v_1\right) \tag{2.30}$$

$$\boldsymbol{r}_v = \frac{1}{\Delta u \Delta v}\left(-\boldsymbol{r}_{11}u_2 + \boldsymbol{r}_{12}u_2 + \boldsymbol{r}_{21}u_1 - \boldsymbol{r}_{22}u_1\right) \tag{2.31}$$

$$\boldsymbol{r}_{uv} = \frac{1}{\Delta u \Delta v}\left(\boldsymbol{r}_{11} - \boldsymbol{r}_{12} - \boldsymbol{r}_{21} + \boldsymbol{r}_{22}\right) \tag{2.32}$$

Obviously, $\boldsymbol{r}_c$ is the position vector of the origin of the local (u, v) co-ordinate system, which can be chosen to be zero. Only one of the vectors $\boldsymbol{r}_u$, $\boldsymbol{r}_v$ and $\boldsymbol{r}_{uv}$ can be zero. If $\boldsymbol{r}_{uv}$ is zero, the bilinear surface degenerates into a rhomboid. If, in addition, $\boldsymbol{r}_u$ and $\boldsymbol{r}_v$ are mutually perpendicular, the rhomboid degenerates into a rectangle. In particular, if $\boldsymbol{r}_{uv}$ is nonzero and is in the plane defined by the vectors $\boldsymbol{r}_u$ and $\boldsymbol{r}_v$, the bilinear surface degenerates into a flat quadrilateral. It can easily be shown that, if $\boldsymbol{r}_u = -v_i\boldsymbol{r}_{uv}$, $i = 1, 2$, namely $\boldsymbol{r}_v = -u_i\boldsymbol{r}_{uv}$, $i = 1, 2$, the bilinear surface degenerates into a trinagle. If we also adopt that $u_i = 0\,(v_i = 0)$, we obtain $\boldsymbol{r}_v = 0\,(\boldsymbol{r}_u = 0)$.

Although a bilinear surface in the general case is curved, all the u and v co-ordinate lines are straight lines, as indicated in Fig. 2.11. This is precisely why such surfaces are termed 'bilinear'. Note that, owing to this property, a bilinear surface cannot be concave or convex, but only inflected (or planar). Therefore, although a curved surface, it can be used for efficient approximation only of those surfaces, or parts of surfaces, which do not have pronounced concave or convex properties.

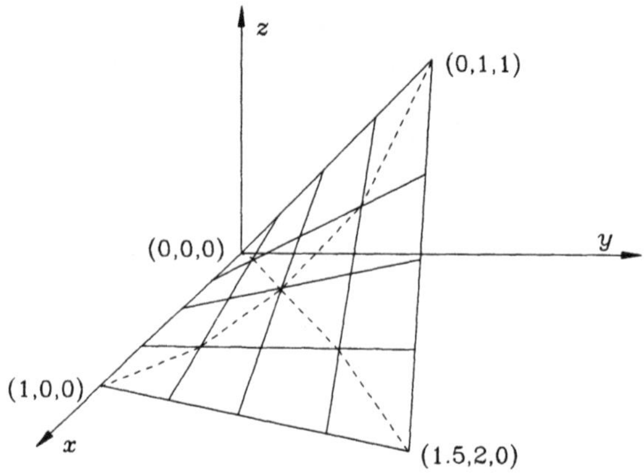

Figure 2.11 Typical bilinear surface

Starting from eqn. 2.28, it is a simple matter to compute any other quantity of interest. For example, the Lamé coefficients are obtained as

$$e_u = \left|\frac{\mathrm{d}\boldsymbol{r}}{\mathrm{d}u}\right| \qquad \frac{\mathrm{d}\boldsymbol{r}}{\mathrm{d}u} = \boldsymbol{r}_u + \boldsymbol{r}_{uv}v \tag{2.33}$$

$$e_v = \left|\frac{\mathrm{d}\boldsymbol{r}}{\mathrm{d}v}\right| \qquad \frac{\mathrm{d}\boldsymbol{r}}{\mathrm{d}v} = \boldsymbol{r}_v + \boldsymbol{r}_{uv}u \tag{2.34}$$

and the unit vectors $\boldsymbol{i}_u$ and $\boldsymbol{i}_v$ are obtained according to eqn. 2.9.

To obtain the simplest algorithm possible, the starting and end u and v coordinates for a bilinear surface will always be chosen to be -1, i.e. $+1$ if not explicitly stated otherwise.

2.7.3 Approximation of surfaces by bilinear surface elements

Partitioning of surfaces into smaller surface elements may be necessary if surfaces are electrically large and if they are curved. Extending the reasoning valid for generalised wires, the number of functions necessary for approximation of surface currents on a surface is roughly proportional to the electrical size of the surface (i.e. to its area normalised with respect to wavelength squared). Most frequently, the surface current over a surface is approximated by a functional series having only a few terms. Therefore the electrical size of the surface may frequently not be large. For example, it is well known that subdomain basis functions cannot be successfully applied if the maximum dimension of the patch side exceeds about 0.1–0.25λ [46, 49]. Entire-domain basis functions, however, can be used up to patch-side lengths as great as one or two wavelengths.

As pointed out in Section 2.5, when a curved surface is approximated by surface elements satisfactory results for the surface-current distribution are obtained only if the approximate surface does not differ excessively from the original surface. As a rough estimate, for electrically small surfaces the maximum deviation of the approximate surface with respect to the original should be compared with some characteristic length describing the surface and should not exceed a certain percentage of this length, e.g. 10% of the maximum linear dimension of the object. For electrically larger surfaces, the maximum deviation should be compared with the operating wavelength, and should not exceed about 1/32nd or at the most 1/16th of the wavelength [82]. In analogy with wires, for electrically small surfaces there is no sense in comparing the deviation with the wavelength, while for electrically larger surfaces the wavelength must be used as a reference.

With these constraints on the maximum permitted electrical size of bilinear surfaces and maximum permitted deviation of the approximate surface with respect to the original, it is possible to construct many procedures for partitioning a surface. As for generalised wires, it is desirable that the partitioning be such that a good approximation of the surface is obtained with a small number of relatively uniform surface elements. However, in this case also there is no simple algorithm which can guarantee in advance the optimum partitioning of a surface in all cases. Therefore experience also plays an important role in obtaining the optimal approximation of arbitrary quadrilaterals by means of bilinear surfaces.

As an example, consider a spherical scatterer modelled exactly by six generalised quadrilaterals, as sketched in Figure 2.5. The approximation of these quadrilaterals by means of bilinear surface elements can be performed starting from the parametric equations of the quadrilaterals (e.g. from eqn. 2.13 for the upper quadrilateral in Figure 2.5). Of course, there are many ways of choosing

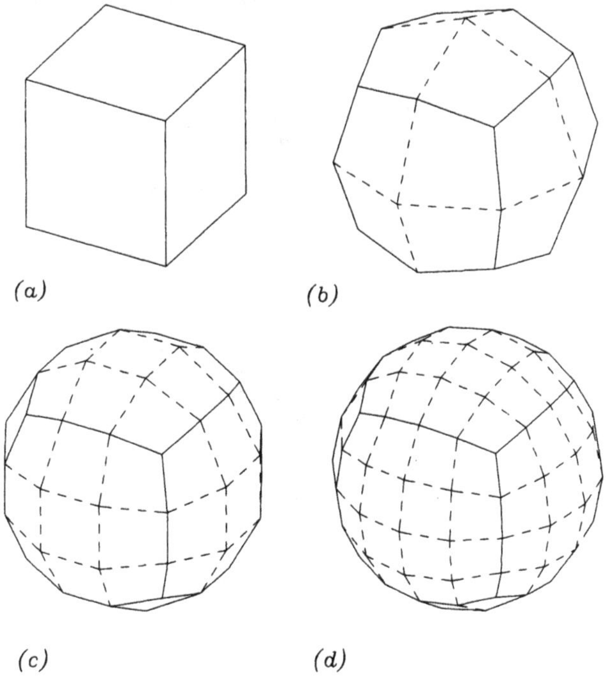

Figure 2.12 *Approximate modelling of a sphere by (a) 6, (b) 24, (c) 54 and (d) 96 bilinear surfaces*

the vertices of these elements. For example, they can be defined by a mesh of uniformly distributed points in the u and v co-ordinates on the scatterer surface. Figures 2.12*a*, *b*, *c* and *d* show four examples of such partitioning, with each quadrilateral on the sphere subdivided into one, four, nine and 16 bilinear surface elements, respectively. It is seen that, in spite of the above-mentioned property of bilinear quadrilaterals that they cannot be concave or convex, the approximation of the sphere seems to be acceptable even with as few as $6 \times 4 = 24$ bilinear surface elements. This intuitive estimate is confirmed quantitatively by the results given in Table 2.2.

Note that this approximation is not optimal for partitioning the quadrilaterals into more than four surface elements. The optimum partitioning can be obtained starting from the above approximations and by optimising the positions of the vertices. Table 2.2 refers to such, slightly changed, positions of the vertices of bilinear surfaces. It gives the following quantities for each number n of subdivisions per quadrilateral side:

(i) the total number of bilinear surfaces N used for the approximation of the sphere
(ii) the maximum deviation $d_{p\max}/a$ between the approximate surface obtained with bilinear surface elements and the original sphere, normalised with respect to the sphere radius
(iii) the maximum electrical radius $a_{\max}/\lambda$ of the sphere for which the maximum deviation is 1/24th of the wavelength
(iv) the corresponding electrical length $l_{\max}/\lambda$ of the largest side among all bilinear surfaces.

Table 2.2 Some quantities used in approximating a sphere of radius a by bilinear surface elements for various numbers of subdivisions n per quadrilateral side

n	N	$d_{p\max}/a$	$a_{\max}/\lambda$	$l_{\max}/\lambda$
1	6	0.423 (0.423)	0.099	0.114
2	24	0.127 (0.113)	0.328	0.253
3	54	0.057 (0.057)	0.718	0.381
4	96	0.033 (0.031)	1.263	0.493

N = total number of bilinear surfaces used for approximation of the sphere
$d_{p\max}/a$ = maximum deviation between the approximate surface obtained with bilinear surface elements and the original sphere, normalised with respect to the sphere radius
$a_{\max}/\lambda$ = maximum electrical radius of the sphere for which the maximum deviation is 1/24th of the wavelength
$l_{\max}/\lambda$ = corresponding electrical length of one side of the bilinear surface
Numbers in parentheses in the third column correspond to the bilinear surfaces approximated each by two flat triangles

The number in parentheses in the third column of Table 2.2 correspond to the bilinear surfaces approximated each by two flat triangles.

It is seen from Table 2.2 that, for electrically small and medium spheres ($a \leqslant 0.3\lambda$), good approximation can be obtained with $N = 24$ surface elements. Naturally, for electrically larger spheres ($a > 0.3\lambda$) more surface elements are needed. Note that, as with a wire-ring scatterer, a_{max}/λ is not proportional to n. It can be easily shown that, for $n \geqslant 2$, this ratio is appproximately proportional to n^2, i.e. as the electrical radius of the sphere increases the number of surface elements should increase approximately as the square root of the sphere radius. This explains why the maximum electrical length of the quadrilateral side increases for larger sphere electrical radii. The values of these lengths show that subdomain basis functions cannot be used for the analysis of larger spheres modelled by the minimum number of surface elements, i.e. additional partitioning is needed. Finally, note that if bilinear surfaces are modelled by two flat triangles approximately the same results are obtained.

2.8 Approximate modelling of geometry of wire-to-plate junctions

As mentioned, surface-patch modelling can be applied not only to plates but also to wires. In this case the geometry of a wire-to-plate junction is modelled in the same way as the geometry of the plate-to-plate junction [83]. However, if we want to use the advantages of modelling wires by means of wire segments, additional attention should be devoted to wire-to-plate junctions.

There are two general approaches to the treatment of wire-to-plate junctions based on modelling wires by means of wire segments. Both use the same tools as for analysis of metallic antennas and scatterers without wire-to-plate junctions. However, one approach introduces specific classes of geometrical element and basis function (known as the 'attachment modes') in the analysis of certain classes of junction. The other approach introduces a specific segmentation technique without introducing any new type of geometrical element or basis function.

The two approaches to the treatment of wire-to-plate junctions result in two geometrical models of the junctions. In what follows these geometrical models will be briefly discussed.

2.8.1 Attachment modes

Consider the arbitrary wire-to-plate junction sketched in Figure 2.13*a*. If the geometry of the junction is known, certain conclusions about the current distribution at the junction can be reached. It is evident that, in the vicinity of such a junction, the quasiradial surface-current component on the plate is dominant, and that the intensity of this component is inversely proportional to the distance from the junction. This means that the current in the vicinity of a wire-to-plate junction can be decomposed into two components: the quasiradial component, flowing from the wire over the generalised frill (see Figure 2.13*b*) and vanishing at the outer frill edge, and the slowly varying component, distributed over the plate. This generalised frill is obviously a new geometrical

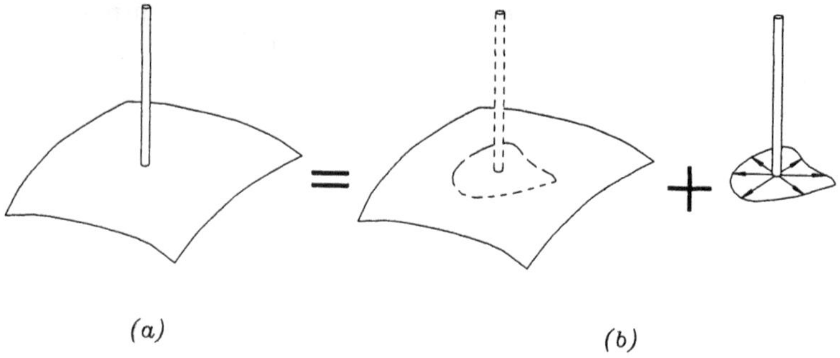

Figure 2.13 *Wire-to-plate junction*

a General form
b Attachment-mode representation

element that should be added to the wire and plate existing at the junction. Such a frill, together with the associated basis function for approximation of the surface current, is referred to as the 'attachment mode'.

Since the circular dependence of the dominant, quasiradial component, is different for different types of wire-to-plate junction, it is clear that many attachment modes can be constructed. The simplest case is obtained when the connection is in the middle of a flat plate [46, 52]. The generalised frill then takes the form of a simple frill, as sketched in Figure 2.14*a*. By numerical experiments it was found that the optimum outer radius of the simple frill is between 0.20 and 0.25 wavelengths [46]. If it is significantly less than 0.20 wavelengths, it turns out to be quite difficult to obtain satisfactory results for the current distribution. It is natural to assume that approximately the same criteria

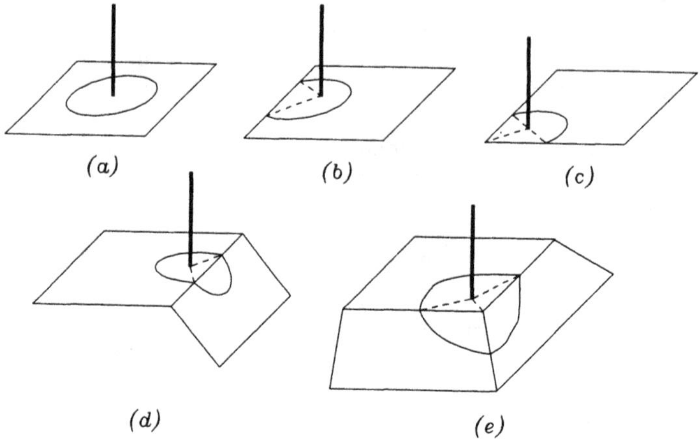

Figure 2.14 *Examples of attachment modes*

a Central
b Edge
c Corner
d Wedge
e Vertex

should be valid also for generalised frills. Therefore, for a junction close to an edge, a corner, a wedge or a vertex, new attachment modes are needed [58, 59], as sketched in Figures 2.14*b*, *c*, *d* and *e*. Note that these generalised frills consist of four types of simpler geometrical element: a complete and a truncated frill segment, and a complete and a truncated frill sector. Neither of these attachment modes can be appled if the radius of the plate curvature is much smaller than the wavelength. In that case other attachment modes are needed.

It is evident that such an approach complicates the analysis significantly. For example, the minimum number of classes of field and impedance integrals which must be evaluated in the analysis is n and n^2, respectively, where n is the number of types of basic surface element used for modelling the geometry. For metallic antennas and scatterers modelled exclusively by means of truncated cones and bilinear surfaces, this number is only two, i.e. $n = 2$ and $n^2 = 4$. However, if attachment modes are used consisting, for example, of four new types of surface element, these numbers become $n = 6$, and $n^2 = 36$.

To simplify this approach, frill segments and sectors can be approximated by flat rectangles (or bilinear surfaces). However, such a method is still somewhat complicated, as will be obvious from the following example. Consider the simple frill, i.e. the central attachment mode, shown in Figure 2.14*a*. Bearing in mind the criteria for surface-patch modelling given in Section 2.7.3 (that the approximate surface should not differ from the original surface in any direction for electrically small surfaces more than about 10% and for electrically larger surfaces for more than about 1/16th–1/32nd of the wavelength), it can be concluded that six bilinear surfaces probably suffice for modelling of a simple frill.

A final possibility in the simplification of modelling of wire-to-plate junctions is elimination of the attachment modes by specific segmentation, as explained in Section 2.8.2.

2.8.2 General localised junction model

Consider the junction of generalised wires and generalised quadrilaterals shown in Figure 2.15*a*. One end of each wire and one side of each quadrilateral is situated in an electrically small domain, which will be referred to as the 'junction domain'. All these ends are assumed to be interconnectd by electrically small pieces of metallic wire or plate (not shown) that are situated in the junction domain.

From the quasistatic analysis, it is evident that the total current flowing out of the junction domain is approximately zero, and that the partial currents flowing through the ends of the wires and the sides of the quadrilaterals in the junction domain do not depend excessively on the shape of the metallic elements interconnecting them. It can therefore be assumed that a satisfactory analysis of such a junction can be obtained if the interconnecting elements are omitted from the geometrical model, provided that the current expansions for the wires and quadrilaterals are adopted in such a manner that the total current flowing out from the junction domain be zero. In other words, the interconnecting elements provide only quasistatic current connections between the ends of wires and sides of quadrilateral plates in the junction domain, and the field due to currents in these elements can be neglected.

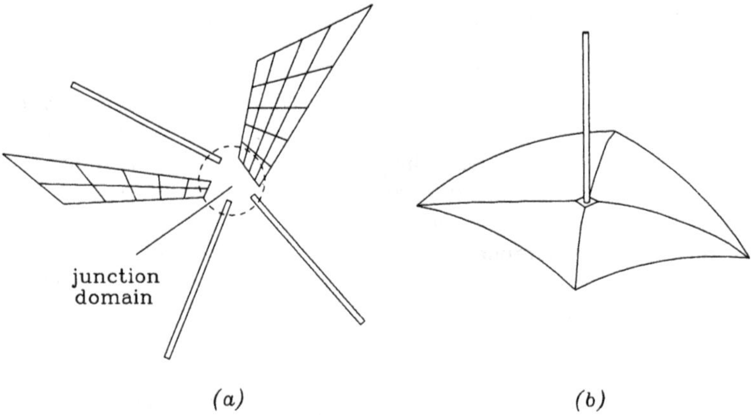

Figure 2.15 Generalised localised junction model

a Definition
b Example

This simple reasoning has far-reaching consequences in the philosophy of modelling of the geometry of metallic antennas and scatterers. It tells us that, in fact, no strict modelling of the interconnections is indispensable, provided that the current-continuity equation is stipulated properly. Specifically, two or more interconnected elements in a model need not be in a physical contact at all, or may be in actual contact at a single point or along only part of their sides. This is

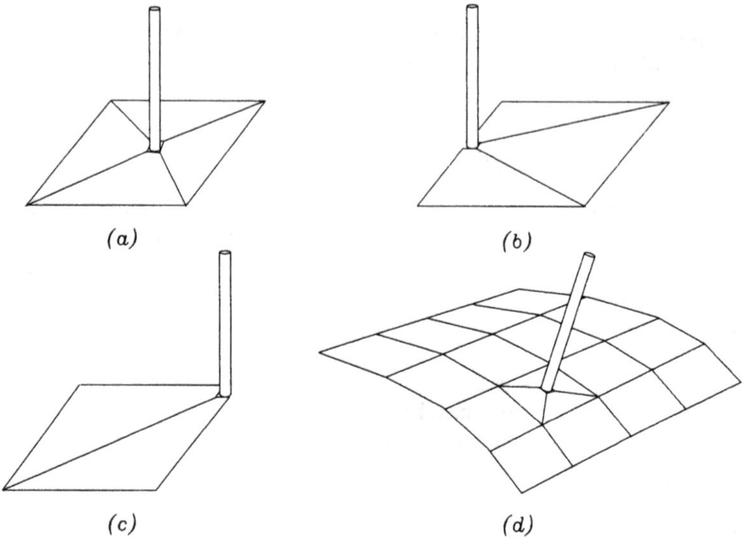

Figure 2.16 Examples of wire-to-plate junction models obtained with truncated cones and bilinear surfaces

a Junction at the plate centre
b Junction at the plate edge
c Junction at the plate vertex
d A curved plate with an attached wire

a very useful conclusion which greatly simplifies modelling of the geometry of junction domains.

Using this simplified model of a junction, it is possible to analyse not only efficiently, but also quite accurately, any wire-to-plate junction. For example, the junction sketched in Figure 2.13*a* can be represented as a junction of one generalised wire and four generalised quadrilaterals, as shown in Figure 2.15*b*. Note that the number of wires or the number of plates forming such a junction can be zero, so that the junction of several wires can be considered as a special case of this junction model. Noting that it has been assumed that the junction domain is electrically small, such a junction model will be referred to as the 'general localised-junction model' [84].

If modelling of geometry is performed by right truncated cones and bilinear surfaces, no attachment modes are needed. For example, the junction of a wire and the central part of a plate can be represented as a junction of a truncated cone and four bilinear surfaces, as shown in Figure 2.16*a*. The junction of a wire and a plate edge can be represented as the junction of a truncated cone and three bilinar surfaces, as shown in Figure 2.16*b*. The junction of a wire and a metallic corner can be represented as the junction of a truncated cone and two bilinear surfaces, as shown in Figure 2.16*c*. Starting from the last two examples it is a simple matter to model the junction of a wire and a metallic wedge, and the junction of a wire and a metallic vertex. Finally, the junction of a wire and a curved plate can also be modelled using this approach, as shown in Figure 2.16*d*.

2.9 Conclusions

Although this Chapter is devoted to modelling of the structure geometry, it is pointed out that the modelling should be carried out bearing in mind the type of approximation of currents over conductor surfaces. In particular, it is very useful to define both the (approximate) surface elements and surface currents over them in the same local co-ordinate system.

The modelling of geometry is carried out with generalised quadrilaterals as basic surface elements which can degenerate into generalised wires and generalised triangles. The exact modelling of surfaces is possible, but approximate modelling is usually both practically sufficient and preferable.

When considering thin wires, approximate modelling can be achieved very efficiently by means of right truncated cones. Surfaces can be approximated in a simple manner by bilinear surface elements.

Particular care is devoted to wire-to-plate junctions. It is demonstrated that the classical attachment modes, which are rather complicated and usually require special attention, can be avoided. Starting from the general localised junction model, a novel partitioning technique for the junction area is proposed. This technique enables the same elements, truncated cones and bilinear surfaces, to be used for modelling of a wire-to-plate junction.

Chapter 3

Approximation of current along generalised wires and over generalised quadrilaterals

3.1 Introduction

It was explained in Chapter 2 how we can approximate wirelike structures by generalised wires and curved surfaces by generalised quadrilaterals. This was the first step in solving the real problem we wish to solve: the determination of current distribution over perfectly conducting bodies in incident time-harmonic electromagnetic fields.

To find the current distribution by means of the method of moments, it is necessary to adopt basis functions in the expansion for the current along generalised wires and for the surface current over generalised quadrilaterals. It was explained in Chapter 2 that it is extremely convenient to describe these currents, i.e. the corresponding basis functions, in the co-ordinate system with respect to which the generalised wire being considered (the generalised quadrilateral) has been described.

As pointed out in Chapter 1, there are three types of basis functions, i.e. current approximations. Briefly, if we define basis functions over entire elements (i.e. over a generalised quadrilateral or along a generalised wire segment), the basis functions are said to be of entire-domain type. If we subdivide the wirelike or surface elements into small subelements and define essentially a single basis function for every subelement, we say that subdomain basis functions are adopted. Finally, if the subdivision of the elements is into only few relatively large subelements, and basis functions are defined for such large generalised wire or surface subelements, the basis functions are said to be of almost entire-domain type. Since almost entire-domain and entire-domain approximation are in essence the same type of approximation, we shall refer to both as entire-domain approximation. A more precise definition of these concepts will be given later in this Chapter.

As mentioned in Chapter 1, subdomain basis functions have been used frequently for the analysis of quite general structures, and have usually been chosen to satisfy the continuity equation at the subelement interconnections and ends or edges [6, 14, 44, 49]. In contrast, until recently only entire-domain basis functions which do not satisfy the continuity equation at the interconnections of surface elements (but do satisfy the conditions at surface-element free edges) have been used. For surfaces without interconnections, no other conditions were necessary [50, 51, 54]. For structures with interconnections, the continuity equation had to be added to the equations obtained by the method of moments [61]. Entire-domain basis functions proved to be efficient and, at least for a given number of unknowns, far superior to subdomain basis functions. In addition, they result in a significantly more accurate near field. Nevertheless, they did not attract wider attention, probably because subdomain basis functions are conceptually and computationally simpler.

This Chapter is aimed at deriving general entire-domain, or almost entire-domain, current expansions (in the above sense), following to some extent the method described in References 66 and 67, but also at explaining some subdomain current expansions. It is shown that basis functions can be defined over the surface of arbitrary generalised quadrilateral surface elements that automatically satisfy the continuity equation along the element interconnections and free edges. This is a very important step, which makes most of the methods proposed so far for the analysis of similar structures special, frequently less accurate, cases of the general method proposed here. Specific polynomial expansions will also be elaborated, and their relative advantages with respect to other possible expansions will be discussed.

3.2 Current expansions along generalised wires

3.2.1 Introduction

Current expansions along generalised wires can in most cases be any of those used for current approximation along thin cylindrical wires, e.g. piecewise-linear expansion, piecewise-sine expansion, polynomial expansion etc. If a generalised wire length is much larger than its average radius, these approximations are certainly appropriate. However, if degenerate forms of generalised wires are considered, such as conical or flat wire ends or the flat ring obtained if we have a sharp transition in wire radius between two interconnected segments, not all thin-wire current expansions seem to be convenient, although they can be used. For example, in the cases mentioned of degenerate generalised wires, both the polynomial approximation and more complicated approximations including terms describing the natural behaviour of the current near edges in question, might be more suitable than other approximations.

Whatever the type of current expansion initially adopted, it is very convenient to transform it to the form in which the current-continuity equation at interconnections and free ends of generalised wires is satisfied automatically. In this manner the number of unknowns can be reduced significantly and, in addition, a more stable solution is obtained. It will be shown that any current expansion can easily be manipulated so that these conditions be satisfied.

3.2.2 Current expansions satisfying continuity equation at ends and interconnections of generalised wires

Suppose we wish to approximate the current distribution along a generalised wire by means of the expansion

$$I(u) = \sum_{i=1}^{n} a_i f_i(u) \qquad u_1 \leqslant u \leqslant u_2 \tag{3.1}$$

where a_i are unknown coefficients to be determined, and $f_i(u)$ are known functions. Let u be the co-ordinate along the adopted wire reference generatrix, u_1 be the co-ordinate of the wire starting point, and u_2 that of the wire end point. We choose the reference direction of the current to be always from the wire starting point towards its end point.

The expansion given in eqn. 3.1 yields the current intensity at the segment starting and end points if we set $u = u_1$ and $u = u_2$. For stipulating the current-continuity equation at the wire starting and end points, however, it is much more convenient to transform the expansion in eqn. 3.1 into the form in which a single term, e.g. I_1 or I_2, represents these current intensities. Therefore we represent the current expansion as

$$I(u) = I_1 g_1(u) + I_2 g_2(u) + I_s(u) \qquad u_1 \leqslant u \leqslant u_2 \tag{3.2}$$

with the condition that

$$I(u_1) = I_1 \quad \text{and} \quad I(u_2) = I_2 \tag{3.3}$$

Current expansions are usually given in the form of eqn. 3.1. To obtain the expansion in the form in eqns. 3.2 and 3.3, we write the expansion in eqn. 3.1 for the starting and end points of the segment and then express any two of the coefficients a_i, for example a_1 and a_2, in terms of I_1 and I_2. The following system of equations is thus obtained:

$$a_1 f_1(u_1) + a_2 f_2(u_1) = I_1 - \sum_{i=3}^{n} a_i f_i(u_1) \tag{3.4}$$

$$a_1 f_1(u_2) + a_2 f_2(u_2) = I_2 - \sum_{i=3}^{n} a_i f_i(u_2) \tag{3.5}$$

Solving these equations for a_1 and a_2 and substituting the expressions for a_1 and a_2 thus obtained into the original expansion in eqn. 3.1, we obtain the following expressions for the functions $g_1(u)$, $g_2(u)$ and $I_s(u)$:

$$g_1(u) = \frac{1}{Q_{12}} \{f_2(u_2) f_1(u) - f_1(u_2) f_2(u)\} \tag{3.6}$$

$$g_2(u) = \frac{1}{Q_{21}} \{f_2(u_1) f_1(u) - f_1(u_1) f_2(u)\} \tag{3.7}$$

$$I_s(u) = \sum_{i=3}^{n} a_i g_i(u) \tag{3.8}$$

where

$$g_i(u) = f_i(u) + \frac{1}{Q_{21}} \{Q_{i2} f_1(u) - Q_{i1} f_2(u)\} \qquad i = 3, \ldots, n \tag{3.9}$$

$$Q_{ij} = f_i(u_1) f_j(u_2) - f_i(u_2) f_j(u_1) \qquad i = 1, 2, \ldots, n \qquad j = 1, 2 \tag{3.10}$$

It is simple to check that $g_1(u_1) = 1$, $g_1(u_2) = 0$, $g_2(u_1) = 0$, $g_2(u_2) = 1$ and $g_i(u_1) = g_i(u_2) = 0$, $i = 3, 4, \ldots, n$, as it should be.

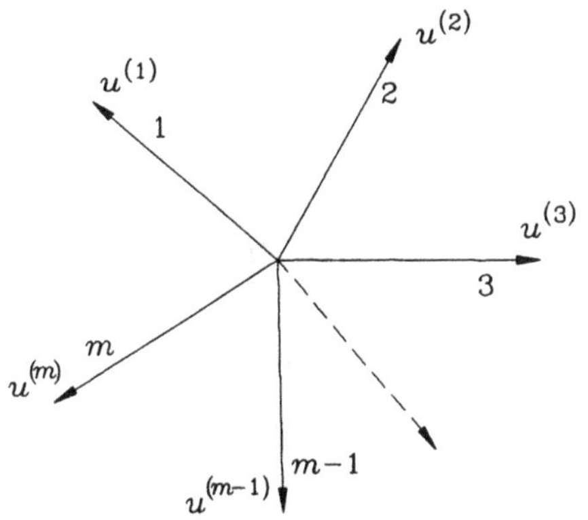

Figure 3.1 Sketch of m generalised wires meeting at a common junction

Let us now explain how we use these expansions to satisfy current-continuity conditions automatically in various cases.

Consider first a free end of a generalised wire (i.e. an end not connected to any other wire). If the free end is the end point of the wire, the continuity equation is satisfied simply by setting $I_2 = 0$. If the free end is the starting point of the wire, the continuity equation is satisfied by setting $I_1 = 0$.

The real usefulness of the expansion in eqn. 3.2 is, however, in determining current expansions for wires meeting at a node (or junction) for which the current-continuity equation at the junction is satisfied automatically for any number of wires. To obtain such an expansion, consider m wires meeting at a junction, as shown in Figure 3.1. Assume that we have found current expansions for individual wires in the form obtained by substituting eqn. 3.8 in eqn. 3.2, i.e. that

$$I^{(k)}(u) = I_1^{(k)} g_1^{(k)}(u) + I_2^{(k)} g_2^{(k)}(u) + \sum_{i=3}^{n_k} a_i^{(k)} g_i^{(k)}(u) \qquad u_1^{(k)} \leqslant u \leqslant u_2^{(k)} \qquad k = 1, 2, \ldots, m \tag{3.11}$$

To simplify the analysis, let the junction be the starting point of all m wires, i.e. let current intensities from the junction into the wires be $I_1^{(k)}$, $k = 1, 2, \ldots, m$. We wish to obtain an expansion which automatically satisfies the current-continuity equation at the junction. Since all junction regions are assumed to be electrically small, the current-continuity equation reads

$$\sum_{k=1}^{m} I^{(k)}\{u_1^{(k)}\} = \sum_{k=1}^{m} I_1^{(k)} = 0 \tag{3.12}$$

To satisfy this equation, the coefficients $I_1^{(k)}$ can be chosen in many ways. The simplest is probably to express one of the coefficients $I_1^{(k)}$, e.g. $I_1^{(1)}$, in terms of the others. Substituting the coefficient $I_1^{(1)}$ thus found into eqn. 3.11 for $k = 1$, we find that the current expansion along the generalised wire labelled 1 is of the form

$$I^{(1)}(u) = -\left\{\sum_{k=2}^{m} I_1^{(k)}\right\} g_1^{(1)}(u) + I_2^{(1)} g_2^{(1)}(u) + \sum_{i=3}^{n_1} a_i^{(1)} g_i^{(1)}(u) \qquad u_1^{(1)} \leqslant u \leqslant u_2^{(1)} \tag{3.13}$$

The expansions for all the other generalised wires meeting at the junction ($k = 2, \ldots, m$) remain as in eqn. 3.11. These are the required expansions for the currents along the m wires that satisfy automatically the current-continuity equation (eqn. 3.12) at the junction.

At the other end of the wires (those not at the junction), only the second term on the right-hand side in eqns. 3.11 and 3.13 is nonzero. Therefore, if the other end of a wire is a free end, we omit this term. If it is not, i.e. if it terminates at some other junction of wires, it must be present. We combine it with the corresponding terms in the expansions along the wires meeting at that other junction, in the manner outlined above, to obtain the expansions for wires meeting at that junction that also satisfy the continuity equation. Only if the wire has both ends free (i.e. is not connected to other wires) does one type of expansion suffice, in the form of the sum in eqn. 3.11.

Note that the first term on the right-hand side in eqn. 3.13, when compared with that in eqn. 3.11, is decomposed into ($m-1$) terms. We can combine each of the terms under the first summation sign on the right-hand side of eqn. 3.13 with the first term in eqn. 3.11, $k = 2, 3, \ldots, m$, and thus obtain the following expansions extending along the generalised wire labelled 1 and the generalised wires labelled 2, 3, . . . , m, that are continuous across the junction:

$$I_d^{(k)}(u) = I_1^{(k)} \begin{cases} -g_1^{(1)}(u) & u_1^{(1)} \leqslant u \leqslant u_2^{(1)} \\ g_1^{(k)}(u) & u_1^{(k)} \leqslant u \leqslant u_2^{(k)} \end{cases} \qquad k = 2, 3, \ldots m \tag{3.14}$$

Obviously, an analogous treatment is possible for the other ends of the generalised wires.

Thus, we can categorise the terms in the proposed current expansions along generalised wires into two types. The first type is zero at both wire ends and does not extend to other wires, possibly connected to the wire considered. Such terms in the expansion we shall therefore term 'singletons'. The terms under the summation sign in eqns. 3.8 and 3.11 and under the second summation sign in eqn. 3.13 are singletons.

The second type of term is characterised by being continuous over the junction of two generalised wires and being zero at the other ends of the two wires. Among many possibilities, they can be obtained by adopting a reference wire at the junction (in the above analysis, the wire labelled 1 has been adopted as the reference wire) and combining part of the reference-wire current at the junction consecutively with the junction currents in the other wires, so that we obtain expansion pairs continuous across the junction. This type of expansion is explained in eqn. 3.14. It can be associated with any pair of interconnected wires, so that a convenient name for such an expansion is a 'doublet'. Whereas

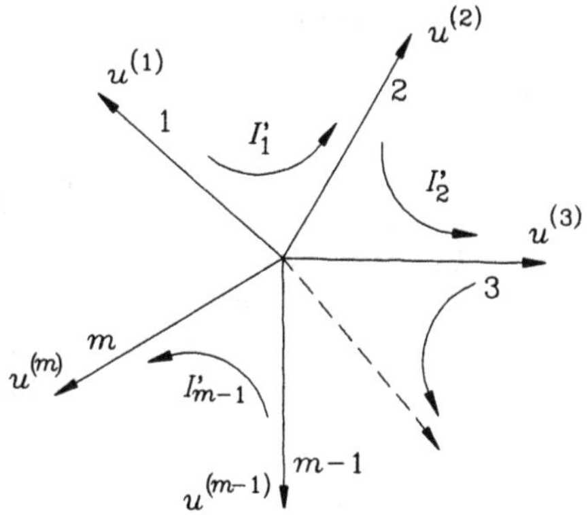

Figure 3.2 'Loop' currents in m generalised wires meeting at a common junction forming various doublets

the number of singletons for any wire is arbitrary, only one doublet can be associated with any two interconnected wires.

Note that singletons can be defined for any generalised wire without knowing the wire interconnections, but that doublets can be defined only if the actual wire interconnections are known and once the choice has been made how to satisfy the current-continuity equation. For example, instead of the convention that the coefficient $I_1^{(1)}$ of the wire labelled 1 (or of the lowest label, or in fact of any wire connected at the junction) be expressed in terms of the coefficients $I_1^{(k)}$ in the current expansions of the other wires meeting at the junction, it is also possible to make other types of choice. For example, consider the junction sketched in Figure 3.2. Imagine 'loop' currents $I_1', \ldots, I_{m-1}'$ as indicated in the Figure. In this case, the $(m-1)$ doublets have exactly these magnitudes at the junction, and are associated with wires 1 and 2, 2 and 3, . . . , $(m-1)$, m (instead of with wires 1 and 2, 1 and 3, . . . , 1 and m):

$$I_d^{(k)}(u) = I_1^{(k)} \left\{ \begin{array}{ll} -g_1^{(k-1)}(u) & u_1^{(k-1)} \leqslant u \leqslant u_2^{(k-1)} \\ g_1^{(k)}(u) & u_1^{(k)} \leqslant u \leqslant u_2^{(k)} \end{array} \right\} \qquad k = 2, 3, \ldots, m \tag{3.15}$$

Finally, only rarely it is possible at all junctions to have the reference

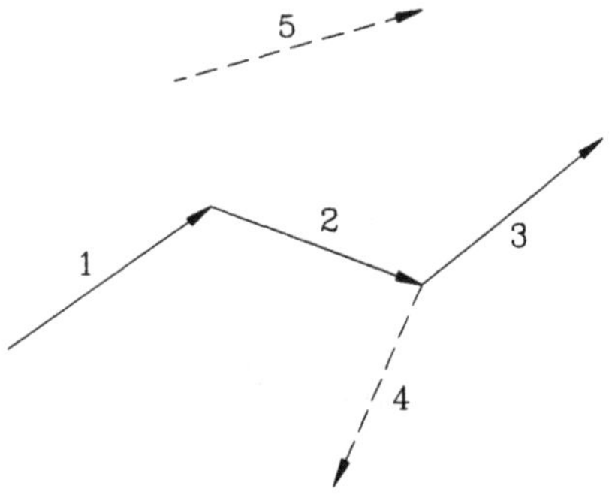

Figure 3.3 Examples of interconnections of generalised wires

directions of all generalised wires meeting at a junction to be from the junction outward. Normally, some will have the opposite reference direction. For such wires, instead of the coefficient I_1, the coefficient I_2 enters the doublets and the sign of the function is changed.

To make the proposed expansions clearer, consider Figure 3.3. It depicts schematically all possible types of connection of generalised wires (including a wire with both ends free). Consider first the series connection of wires only (omitting the two wires indicated by broken lines). The expansions which we need to use for the three wires, with respect to the reference directions of the wires indicated in the Figure and adopting the wire with the lowest label as the reference wire, are as follows. For the first wire,

$$I^{(1)}(u) = I_2^{(1)} g_2^{(1)}(u) + \sum_{i=3}^{n_1} a_i^{(1)} g_i^{(1)}(u) \qquad u_1^{(1)} \leqslant u \leqslant u_2^{(1)} \tag{3.16}$$

because its starting point is a free end, so that $I_1^{(1)}$ must be set to zero. For the second wire all the terms in the expansion in eqn. 3.11 must be present, but note that $I_1^{(2)} = I_2^{(1)}$:

$$I^{(2)}(u) = I_2^{(1)} g_1^{(2)}(u) + I_2^{(2)} g_2^{(2)}(u) + \sum_{i=3}^{n_2} a_i^{(2)} g_i^{(2)}(u) \qquad u_1^{(2)} \leqslant u \leqslant u_2^{(2)} \tag{3.17}$$

The first term on the right-hand side of eqn. 3.16 and the first term on the right-hand side of the last equation form the doublet. For the third wire we have that $I_1^{(3)} = I_2^{(2)}$, and $I_2^{(3)} = 0$, because the end of the third wire is a free end. Thus

$$I^{(3)}(u) = I_2^{(2)} g_1^{(3)}(u) + \sum_{i=3}^{n_3} a_i^{(3)} g_i^{(3)}(u) \qquad u_1^{(3)} \leqslant u \leqslant u_2^{(3)} \tag{3.18}$$

The second term on the right-hand side of eqn. 3.17 and the first term on the right-hand side of eqn. 3.18 form a doublet. Note that the expansions in eqns. 3.16–3.18 automatically satisfy the current-continuity conditions at free ends and interconnections of the wires and that, in addition, the number of unknowns is reduced.

Consider now the expansions if all the indicated wires in Figure 3.3 are present. The expansion for the first wire remains the same as before. However, since

$$-I_2^{(2)} + I_1^{(3)} + I_1^{(4)} = 0 \tag{3.19}$$

we can express one of these coefficients in terms of two others, e.g. we can write $I_2^{(2)} = I_1^{(3)} + I_1^{(4)}$. Therefore the expansion for the second wire in this case becomes

$$\begin{aligned} I^{(2)}(u) = {} & I_2^{(1)} g_1^{(2)}(u) + \{I_1^{(3)} + I_1^{(4)}\} g_2^{(2)}(u) \\ & + \sum_{i=3}^{n_2} a_i^{(2)} g_i^{(2)}(u) \qquad u_1^{(2)} \leqslant u \leqslant u_2^{(2)} \end{aligned} \tag{3.20}$$

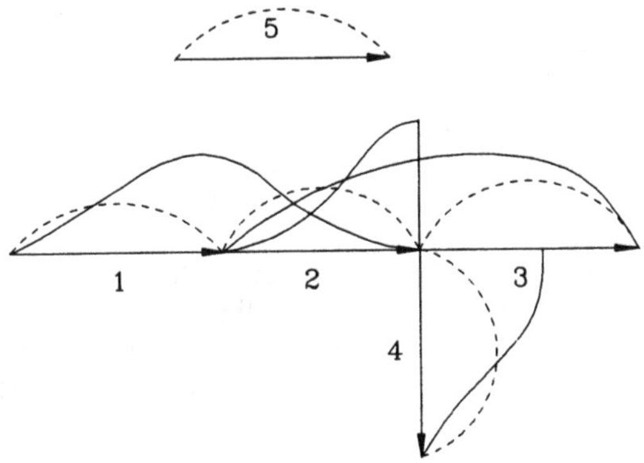

Figure 3.4 Examples of singletons and doublets along generalised wires

— — — — singletons
———— doublets

For the third wire in this case we have

$$I^{(3)}(u) = I_1^{(3)}g_1^{(3)}(u) + \sum_{i=3}^{n_3} a_i^{(3)}g_i^{(3)}(u) \qquad u_1^{(3)} \leqslant u \leqslant u_2^{(3)} \tag{3.21}$$

A similar expansion is valid for the fourth wire:

$$I^{(4)}(u) = I_1^{(4)}g_1^{(4)}(u) + \sum_{i=3}^{n_4} a_i^{(4)}g_i^{(4)}(u) \qquad u_1^{(4)} \leqslant u \leqslant u_2^{(4)} \tag{3.22}$$

Note that the first term in the curly brackets on the right-hand side of eqn. 3.20 and the first term on the right-hand side of eqn. 3.21, as well as the second term in the curly brackets on the right-hand side of eqn. 3.20 and the first term on the right-hand side of eqn. 3.22, form doublets.

Finally, for the last (fifth) wire, which has both ends free, the expansion is of the form

$$I^{(5)}(u) = \sum_{i=3}^{n_5} a_i^{(5)}g_i^{(5)}(u) \qquad u_1^{(5)} \leqslant u \leqslant u_2^{(5)} \tag{3.23}$$

All terms in this expansion are singletons.

As an illustration, Figure 3.4 depicts possible singleton and doublet terms of the expansions for all the wires in Figure 3.3. For convenience, the wires in Figure 3.4 are not positioned as in Figure 3.3, but are interconnected in the same way.

3.2.3 Some entire-domain current expansions along generalised wires

Theoretically, the expansions introduced in Section 3.2.2 can be used with any generalised wire length, but lower-order expansions can be used with smaller lengths. In addition, it is always desirable to adopt current expansions which result in the lowest number of unknowns, permitting accurate current-distribution representation. It is also important that the current expansions adopted allow us to use simplest possible numerical integrations of the integrals to be evaluated.

It was explained in Chapter 1 and repeated in the Introduction to this Chapter that we deal with three types of expansions: entire-domain, almost-entire-domain and subdomain. We shall now see that a precise distinction between the three cannot be made without taking into account, at least implicitly (for example by means of the adopted order of current approximation), the electrical length of the generalised wire considered.

For generalised wires, we shall adopt the following definition: if the generalised wire length is such that we need for the entire wire length to adopt at least the 3rd-order approximation ($n = 3$), we shall say that the expansion is an entire-domain expansion. If we subdivide the generalised wire into subsegments, but still use at least the 3rd-order approximation along at least one of the subsegments, we shall term such an expansion almost-entire-domain expansion. Finally, if the subdivision of the generalised wire is such that, along all subsegments, we use the lowest-order approximation possible that guarantees current continuity from one subsegment to the next (i.e. $n = 2$), we say that current expansion is a subdomain expansion. Note that the definition of the almost-entire-domain expansion is essentially the same as that of the entire-domain expansion. Therefore, as already mentioned in the Introduction to this Chapter, we shall use the term 'entire-domain expansion' for both.

Bearing in mind that wire structures we consider are completely general, it is advisable to adopt the most flexible current expansion possible. In the authors' experience, the polynomial expansion appears to be very convenient for electrical wire lengths not exceeding a few wavelengths. For longer wires, entire-domain current expansion in the form of few trigonometric terms plus a corrective polynomial appears to be very efficient [81]. This will be considered in more detail later in this Section.

By adopting the functions f_i in eqn. 3.1 in the form

$$f_i(u) = (u/L)^{i-1} \qquad u_1 \leqslant u \leqslant u_2, \qquad i = 1, 2, \ldots, n \tag{3.24}$$

the basic general form of the polynomial current expansion can be written as

$$I(u) = \sum_{i=1}^{n} a_i \, (u/L)^{i-1} \qquad u_1 \leqslant u \leqslant u_2 \tag{3.25}$$

In most instances the starting co-ordinate u_1 is chosen to be zero, the end co-ordinate u_2 to be equal to the length of the wire segment and the normalisation coefficient L to be 1. However, such a choice may not be optimal if one bears in mind the following three natural requirements: maximum simplicity of the final expansions (with singleton and doublet terms), the lowest number of algebraic operations in computing $I(u)$, and the simplest modification possible of the expansion if symmetry of current is applicable.

To demonstrate this, let us consider the following examples:

$L = 1,\ u_1 = 0,\ u_2 = l$

$L = l,\ u_1 = 0,\ u_2 = l$

$L = l,\ u_1 = -l,\ u_2 = l$

$L = 1,\ u_1 = -1,\ u_2 = 1.$

Bearing in mind eqn. 3.11 and eqns. 3.6–3.10, the polynomial expansions in these four cases can be written in the following forms, respectively:

$$I(u) = I_1(1 - u/l) + I_2(u/l) + \sum_{i=3}^{n} a_i l^{i-1}\{(u/l)^{i-1} - u/l\} \tag{3.26}$$

$$I(u) = I_1(1 - u/l) + I_2(u/l) + \sum_{i=3}^{n} a_i\{(u/l)^{i-1} - u/l\} \tag{3.27}$$

$$I(u) = \tfrac{1}{2} I_1(1 - u/l) + \tfrac{1}{2} I_2(1 + u/l) + \sum_{\substack{i=3\\(2)}}^{n} a_i\{(u/l)^{i-1} - 1\} + \sum_{\substack{i=4\\(2)}}^{n} a_i\{(u/l)^{i-1} - u/l\} \tag{3.28}$$

$$I(u) = I_1\,\frac{1-u}{2} + I_2\,\frac{1+u}{2} + \sum_{\substack{i=3\\(2)}}^{n} a_i(u^{i-1} - 1) + \sum_{\substack{i=4\\(2)}}^{n} a_i(u^{i-1} - u) \tag{3.29}$$

The symbol (2) below the two summation signs in the last two equations means that the sums should be taken with a step of 2, i.e. $i = 3, 5, \ldots$ in the first sum, and $i = 4, 6, \ldots$ in the second.

First, note that all these (and other) entire-domain expansions have the following useful property: if we wish to increase the order of approximation (i.e. the accuracy of approximation), we need only add one or more higher-degree basis functions. If desired, it is even possible to make use of integrals already computed in doing this. This is not the case with subdomain expansions to increase the accuracy of approximation of which we must subdivide the segment further and no possibility normally exists of using any of the computations made in the preceding step. Secondly, note that the last two expansions enable account to be taken of symmetry or antisymmetry of the current along the wire with respect to the central segment point simply by omitting the even (odd) terms in the two expansions. Finally, note that the last expansion can be considered the simplest of the four in the above sense. Therefore we shall mainly use that expansion.

For the conditionally optimal expansion in eqn. 3.29, the singleton expansion for any wire is seen to be of the form

$$I_s(u) = \sum_{i=3}^{n_s} a_i \begin{Bmatrix} u^{i-1} - 1, \; i\,\text{odd} \\ u^{i-1} - u, \; i\,\text{even} \end{Bmatrix} \tag{3.30}$$

The $(m-1)$ doublets for m wires meeting at an interconnection, with the starting point of all branches at the interconnection [i.e. at $u_1^{(k)} = -1$, $k = 1, 2, \ldots, m$], in that expansion are seen to be

$$I_d^{(k)}(u) = I_1^{(k)} \begin{Bmatrix} \dfrac{u^{(1)} - 1}{2} & -1 \leqslant u^{(1)} \leqslant 1 \\ \dfrac{1 - u^{(k)}}{2} & -1 \leqslant u^{(k)} \leqslant 1 \end{Bmatrix} \qquad k = 2, 3, \ldots, m \tag{3.31}$$

We shall see in later Chapters that numerical results indicate that the polynomial-type expansions result in accurate current distributions with only three to four unknowns per wavelength.

The efficiency of simple polynomial expansions in thin-wire analysis can be improved by adding trigonometric terms. For example, the functions f_i, $i = 1, 2, \ldots, n$, in eqn. 3.1 can be adopted in the form

$$f_i(u) = u^{i-1} \qquad -1 \leqslant u \leqslant 1 \qquad i = 1, 2, \ldots, (n-2) \tag{3.32}$$

$$f_{n-1}(u) = \sin klu \qquad f_n(u) = \cos klu \qquad -1 \leqslant u \leqslant 1 \tag{3.33}$$

where the length of the wire segment is $2l$. The coefficient k is normally chosen to be the phase coefficient of the medium at the frequency considered, $k = \beta = \omega\sqrt{(\epsilon\mu)}$, because the trigonometric functions in eqns. 3.33 are the solutions of the homogeneous part of the equation for current distribution along a straight, infinitely thin wire. Obviously, they represent the dominant part of the current along thin wires of arbitrary shape.

Starting from eqns. 3.6–3.10, the combined polynomial/trigonometric expansion can be derived in the form of eqn. 3.11. However, the form of the expansion is different for $n = 2$, $n = 3$ and $n > 3$. For $n = 2$,

$$I(u) = I_1 \frac{\sin\{kl(1-u)\}}{\sin(2kl)} + I_2 \frac{\sin\{kl(1+u)\}}{\sin(2kl)} \tag{3.34}$$

For $n = 3$,

$$I(u) = \frac{1}{2} I_1 \left\{1 - \frac{\sin(klu)}{\sin(kl)}\right\} + \frac{1}{2} I_2 \left\{1 + \frac{\sin(klu)}{\sin(kl)}\right\} + a_3\{\cos(klu) - \cos(kl)\} \tag{3.35}$$

and for $n > 3$,

$$I(u) = I_1 \frac{1-u}{2} + I_2 \frac{1+u}{2} + \sum_{\substack{i=3 \\ (2)}}^{n-2} a_i(u^{i-1} - 1) + \sum_{\substack{i=4 \\ (2)}}^{n-2} a_i(u^{i-1} - u) + a_{n-1}\{\sin(klu) - u\sin(kl)\} + a_n\{\cos(klu) - \cos(kl)\} \tag{3.36}$$

Note that for $n > 3$ this expansion is the same as the polynomial expansion in eqn. 3.29 except for the last two terms (those of highest degrees) which are replaced by sine and cosine functions.

Other combined expansions can be obtained if some other pair of power functions is replaced by trigonometric functions. Some of the power functions can even be omitted. Numerical results have indicated that approximately the same results are obtained if, in the combined expansion, the constant and the linear terms are omitted. In that case the initial basis functions $f_i(u)$, $i = 1, 2, \ldots, n$, are adopted in the form

$$f_i(u) = u^{i+1} \qquad -1 \leqslant u \leqslant 1 \qquad i = 1, 2, \ldots, (n-2) \tag{3.37}$$

$$f_{n-1}(u) = \sin klu \qquad f_n(u) = \cos klu \qquad -1 \leqslant u \leqslant 1 \tag{3.38}$$

The combined polynomial/trigonometric expansion in the form of eqn. 3.11 based on the above initial basis functions also has different forms for $n = 2$, $n = 3$ and $n > 3$. For $n = 2$, the expansion is given by eqn. 3.34. For $n = 3$, the expansion reads

$$\begin{aligned} I(u) = & \frac{1}{2} I_1 \left\{ u^2 - \frac{\sin(klu)}{\sin(kl)} \right\} + \frac{1}{2} I_2 \left\{ u^2 + \frac{\sin(klu)}{\sin(kl)} \right\} \\ & + a_3 \{\cos(klu) - u^2 \cos(kl)\} \end{aligned} \tag{3.39}$$

and for $n > 3$

$$\begin{aligned} I(u) = & I_1 \frac{u^2 - u^3}{2} + I_2 \frac{u^2 + u^3}{2} \\ & + \sum_{\substack{i=3 \\ (2)}}^{n-2} a_i (u^{i-1} - u^2) + \sum_{\substack{i=4 \\ (2)}}^{n-2} a_i (u^{i-1} - u^3) \\ & + a_{n-1} \{\sin(klu) - u^3 \sin(kl)\} + a_n \{\cos(klu) - u^2 \cos(kl)\} \end{aligned} \tag{3.40}$$

Note that the combined polynomial/trigonometric expansions are significantly more complicated for implementation in thin-wire analysis than simple polynomial expansions. Compared with simple polynomial expansions, they are defined by three different expressions instead of only one, and each of these expressions is more complicated than the corresponding expression in the simple polynomial expansion. In addition, the application of combined expansions in thin-wire analysis involves evaluation of the potential, field and impedance integrals which involve not only polynomials but also trigonometric functions. For electrically longer wire segments, however, they permit very accurate current representation by a fairly small number of terms. From numerical experiments it has been found that the combined expansion in eqns. 3.37–3.40 of the order $(n-2)$ is approximately equivalent to the combined expansion in eqns. 3.32–3.36 of the order n. It has also been found that, for electrically longer wires, the latter combined expansion yields an accurate current distribution with only one to two unknowns per wavelength.

In conclusion, the following recommendations can be made on the choice of the basis functions in the analysis of thin wires: the entire-domain polynomial expansion should generally be used for electrically shorter segments, and almost entire-domain expansions for electrically longer segments. Only if more accurate analysis of electrically longer wires with very small numbers of unknowns per wavelength is of interest, should the combined polynomial/trigonometric expansion be considered as a possibility.

Of course, it is possible to devise many other types of basis function. However, the two simple types mentioned have been found by the authors to be so flexible that no search for more efficient basis functions appeared to be justified.

3.2.4 Subdomain current expansions

We have defined subdomain current expansions as those obtained on a segmented generalised wire and with the lowest approximation of current ($n=2$) along all subsegments which guarantees current continuity from one subsegment to the next. According to eqn. 3.11 it has the form

$$I(u) = I_1 g_1(u) + I_2 g_2(u) \tag{3.41}$$

with $g_1(u_1) = 1$, $g_1(u_2) = 0$, $g_2(u_1) = 0$ and $g_2(u_2) = 1$.

Note that this expansion requires one unknown coefficient per junction between subsegments of a single segmented wire. For example, consider a generalised wire approximated by s subsegments. Let the beginning of the first subsegment coincide with the starting point of the wire, and the beginning of all other subsegments coincide with the end of the preceding subsegment. The current expansion along the first subsegment is of the form of the second term on the right-hand side of eqn. 3.41:

$$I^{(1)}(u) = I_2^{(1)} g_2^{(1)}(u) \tag{3.42}$$

because the current intensity at the free starting point of the first segment must be zero. The current expansion along the second subsegment is of the form in eqn. 3.41, but note that, because of current continuity, $I_1^{(2)} = I_2^{(1)}$:

$$I^{(2)}(u) = I_2^{(1)} g_1^{(2)}(u) + I_2^{(2)} g_2^{(2)}(u) \tag{3.43}$$

A similar expansion is valid for any subsegment except the first and the last:

$$I^{(j)}(u) = I_2^{(j-1)} g_1^{(j)}(u) + I_2^{(j)} g_2^{(j)}(u) \qquad j = 2, 3, \ldots, (s-1) \tag{3.44}$$

Along the last segment,

$$I^{(s)}(u) = I_2^{(s-1)} g_1^{(s)}(u) \tag{3.45}$$

because at its end current intensity is zero.

The most important subdomain approximations are piecewise-constant approximation, piecewise-linear approximation, and piecewise-sine approximation. In the last two cases it is possible to construct basis functions which automatically satisfy the continuity equation at the ends and interconnections of generalised wires. In that case so-called triangular (or triangle) functions (linear doublets) and sine doublets are obtained.

The triangular function is a special case of a polynomial expansion ($n = 2$) described in Section 3.2.3; it is completely defined by eqn. 3.31. It is well known that such an expansion requires about ten unknowns per wavelength to obtain results of acceptable accuracy for current distribution.

The sine doublet is a special case of a combined polynomial/trigonometric expansion ($n = 2$) described in Section 3.2.3. For m wires meeting at an interconnection, with the starting point of all the wires at the interconnection, i.e. at $u_1^{(k)} = -1$, $k = 1, 2, \ldots, m$, the $(m-1)$ sine doublets are

$$I_d^{(k)}(u) = I_1^{(k)} \left\{ \begin{array}{ll} \dfrac{\sin[kl^{(1)}\{u^{(1)} - 1\}]}{\sin\{2kl^{(1)}\}} & -1 \leqslant u^{(1)} \leqslant 1 \\ \dfrac{\sin[kl^{(k)}\{1 - u^{(k)}\}]}{\sin\{2kl^{(k)}\}} & -1 \leqslant u^{(k)} \leqslant 1 \end{array} \right\} \quad k = 2, 3, \ldots, m \quad (3.46)$$

where $u^{(k)}$ is the local (nondimensional) co-ordinate, and the length of the kth segment is $2l^{(k)}$. It is well known that this expansion requires about four unknowns per wavelength to give meaningful results for current distribution. However, to obtain relatively accurate results, particularly for electrically thicker wires, many more unknowns per wavelength are needed.

Thus, both types of subdomain expansion require significantly more unknowns per wavelength than do entire-domain expansions to generate current representations of approximately the same accuracy. It is frequently necessary to analyse wire structures where the total electrical length of the segments amounts to many wavelengths. Subdomain expansions then become useless even if large computers are used for the structure analysis. This book is aimed at presenting a method for the analysis of wire structures with low memory requirements. Therefore the entire-domain-expansion philosophy has been adopted, and no further attention will be devoted to subdomain expansions.

3.2.5 Quasistatic treatment of ends and interconnections of generalised wires

For generalised wires, particularly for approximately resonant structures, accurate treatment of ends of wires may contribute significantly to the overall accuracy of the results. Several papers which deal with that problem [33, 66, 80] solve it by essentially quasistatic analysis. However, with the approximation of generalised wires by truncated cones, the problem of wire ends can also be solved without recourse to quasistatic analysis, since wire ends can be modelled fairly accurately by means of truncated cones. Therefore wire ends can if required be modelled in the same way as the rest of the generalised wire structure. In that case the polynomial expansion containing one singleton suffices for conical and flat ends. Since one doublet is needed for the interconnection of the cylindrical wire section and the conical or flat wire end, it can be concluded that relatively accurate treatment of the wire end requires two additional unknowns.

By observing the exact quasistatic relation at the conical or flat wire end, it is possible to treat wire ends accurately with only one additional unknown. Consider the generalised cone (a cone with arbitrary cross-section) shown in Figure 3.5. It is known that surface-charge distribution in the vicinity of the apex of the cone can be expressed as [85]

$$\rho_s(u, v) = a(v)(u - u_1)^b \qquad b > -1 \qquad (3.47)$$

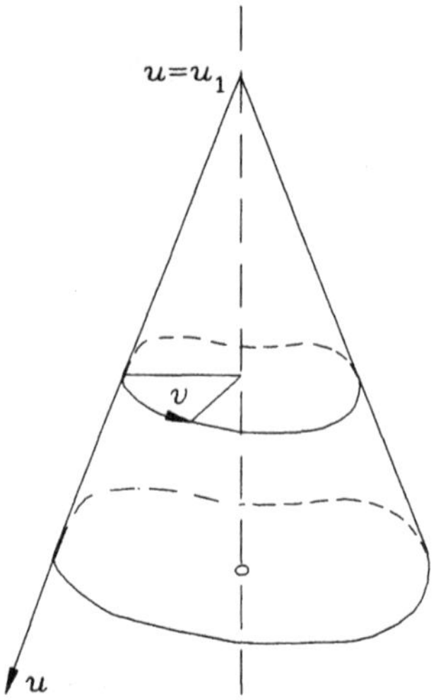

Figure 3.5 A generalised cone

where u is the local co-ordinate along the cone generatrix, v is the local angle co-ordinate, u_1 is the u co-ordinate of the apex, and the function $a(v)$ and the coefficient b depend on the shape of the cone. From this expression it follows that the charge per unit length of the cone generatrix depends on the distance u from the apex as $(u - u_1)^{b+1}$, $b > -1$, i.e. that the charge per unit length is zero at the cone apex. Bearing in mind the continuity equation for line currents, the following condition must be satisfied by the current at the cone apex:

$$\left.\frac{\mathrm{d}I(u)}{\mathrm{d}u}\right|_{u=u_1} = 0 \tag{3.48}$$

By substituting the expression for $I(u)$ in eqn. 3.1 into eqn. 3.48, it is seen that the simplest way of satisfying that equation is to adopt the initial basis functions f_i that satisfy the condition

$$\left.\frac{\mathrm{d}f_i(u)}{\mathrm{d}u}\right|_{u=u_1} = 0 \tag{3.49}$$

For example, for polynomial expansions the functions f_i can be adopted in the form given in eqn. 3.24, with $u_1 = 0$, and omitting the term for $i = 2$. However, note that the polynomial expansion for the conical wire end is then different from that for the cylindrical part of the wire, which makes the algorithm for thin-wire analysis more complicated.

The condition in eqn. 3.48 can also be fulfilled starting from the polynomial expansion along the cylindrical wire section. For example, since the current intensity at the cone apex is zero and a 3rd-order expansion suffices for the conical end, the polynomial expansion in eqn. 3.29 can be written in the form

$$I(u) = I_2\,\frac{1+u}{2} + a_3(u^2 - 1) \tag{3.50}$$

In this case the condition in eqn. 3.48 is satisfied by adopting a_3 as

$$a_3 = -\tfrac{1}{4} I_2 \tag{3.51}$$

Of course, it is possible also to take the formula in eqn. 3.47 into account by introducing specific basis functions. However, such an expansion would be inconvenient, because it is relatively complicated and can only be valid at the wire end. In addition, numerical results indicated that such a sophisticated treatment of the ends does not contribute significantly to overall accuracy of current-distribution calculation.

For electrically thin generalised wires, the analysis of their interconnections could, in principle, also be solved approximately by quasistatic analysis. However, this is usually extremely complicated even from the quasistatic point of view. In addition, every interconnection is a specific problem, and a general approach is quite intricate. Fortunately, charge density at an interconnection is usually negligible, so that accurate modelling of interconnections is not usually critical. In some simple cases which it was possible to analyse (e.g. interconnection of two coaxial conductors of different radii), numerical results indicated that stipulating quasistatic relations may even decrease the accuracy of the complete solution. Therefore no further attention will be devoted to modelling of wire interconnections other than in Section 2.5. Generally speaking, it was found numerically that, if one attempts to obtain very accurate current and charge distributions in one, usually small, region, this may degrade overall accuracy of current-distribution calculation.

3.3 Current expansions over generalised quadrilaterals

3.3.1 Introduction

The definition of basis functions for surface currents over generalised quadrilaterals have certain similarities with that for currents along generalised wires. However, while for electrically thin or axially symmetrical wires it was possible to consider the current as a function of the longitudinal wire co-ordinate only, this is not the case for generalised quadrilaterals. This complicates significantly the equations expressing the continuity conditions along interconnections of two quadrilaterals and requires special attention.

The concepts of singletons and doublets, described in Section 3.2 for generalised wires, under certain conditions, are also possible to define for generalised quadrilaterals. Therefore in many respects the presentation in this Section leans on that in Section 3.2.

3.3.2 General considerations

Consider a metallic body in a vacuum and assume that we have approximated the body by a number of generalised quadrilaterals. We assume that all the quadrilaterals are defined by means of two curvilinear co-ordinates u and v, as explained in Chapter 2. Such a quadrilateral is shown in Figure 3.6. Let the body be situated in a time-harmonic incident electromagnetic field of angular

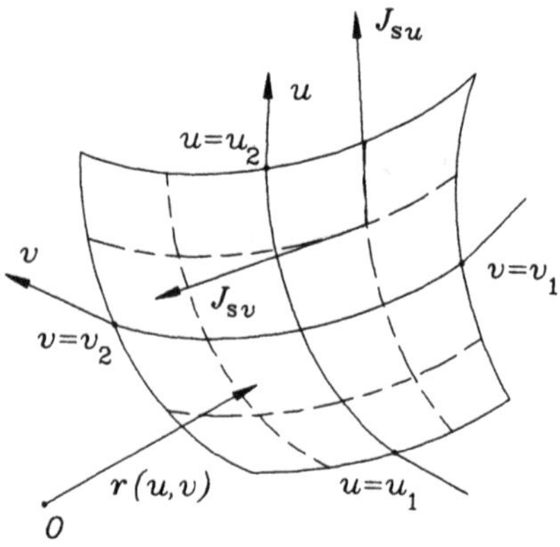

Figure 3.6 A generalised curved quadrilateral

frequency ω, and let the surface-current and surface-charge density over the body be $\boldsymbol{J}_s$ and ρ_s, respectively.

The current over the surface of a curved generalised quadrilateral can be represented at all points of the quadrilateral by means of its u and v components. We have seen, however, that quadrilaterals can be of various degenerate forms. One example is shown in Figure 3.7, in which a degenerate quadrilateral is used to approximate a tubular surface. Another example is sketched in Figure 3.8, in which a curvilinear triangle is obtained from Figure 3.6 if the two lower vertices of the quadrilateral are imagined to overlap. It is also necessary to check in such cases whether representation of the surface-current-density vector is possible by means of its u and v components.

Consider first the case shown in Figure 3.7. The u component of $\boldsymbol{J}_s$ is obviously easily defined. However, in this case there are no other problems concerning representation of the surface-current-density vector by means of its u and v coordinates.

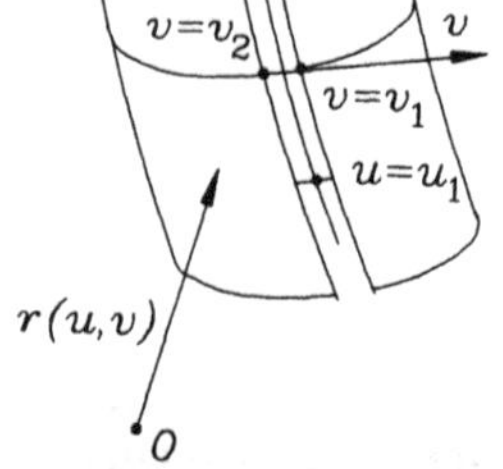

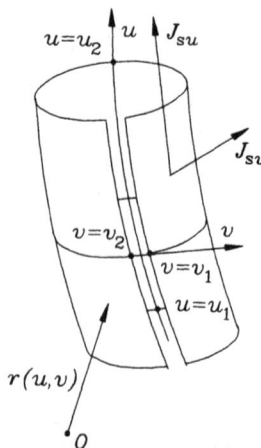

Figure 3.7 Approximation of a cylindrical segment by a degenerate generalised quadrilateral

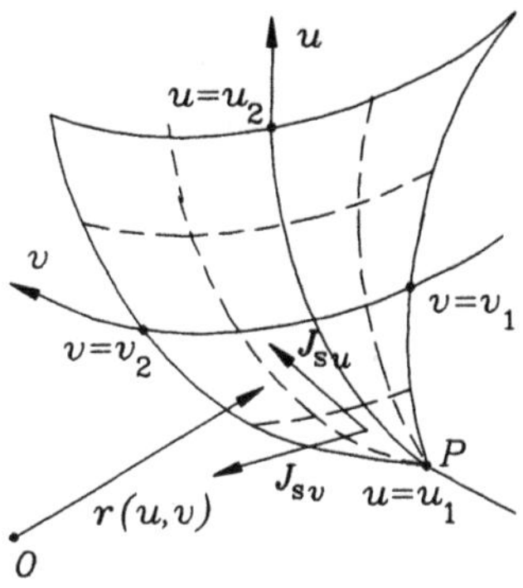

Figure 3.8 Generalised curved triangle as a degenerate case of a curved quadrilateral

Consider now the case of the curved triangle shown in Figure 3.8. Note that in some cases the approximation of a surface can be performed more easily with a few triangles added to the quadrilaterals, so that this is an important degenerate case of generalised quadrilaterals. However, it was briefly explained in Section 2.3.2 that the surface current cannot be represented at all points of the triangle in terms of its u and v components alone. The singular point is the point P in Figure 3.8, at which the u co-ordinate lines intersect (the point $u = u_1$), so that the unit vector $\boldsymbol{i}_u$ at that point is not uniquely defined. Therefore the u component of the current density at such a point can only be zero, which cannot be true except in special cases. Although the v component of the surface-current-density vector at that point is defined uniquely, this is not sufficient to define a vector quantity tangential to the triangle which can otherwise be in any direction.

A remedy is to represent the surface-current-density vector over triangles by means of three components, all three tangential to the triangle at all points. [In fact, in this manner we represent the current-density vector in space, which is actually the case; however, because we have defined the local (u, v) co-ordinate system, it was possible to represent the current-density vector over a quadrilateral by means of only two components.] The third component is, of course, arbitrary. What we wish, however, is to connect it also with the adopted surface element and a local co-ordinate system defined on it. This can be done in the following manner.

Consider first again a quadrilateral element. The choice of the u and v co-ordinates is arbitrary, provided that, as required above, the co-ordinate lines coincide with the sides of the quadrilateral. The co-ordinate system is a local system, with no specific rule dictating which particular co-ordinate of the two should be the u co-ordinate, and which the v co-ordinate. We therefore can imagine both surface-current-density components to be the components labelled in the same manner, for example as the u components, but in two local co-ordinate systems in which the co-ordinates are interchanged and the quadrilaterals overlap, as in Figure 3.9.

Returning now to a triangular element, we can, in analogy, choose all three components of the surface-current-density vector to be the u components in three local co-ordinate systems on the same element, as in Figure 3.10, and imagine that the three triangles overlap. Note, however, that while for a quadrilateral

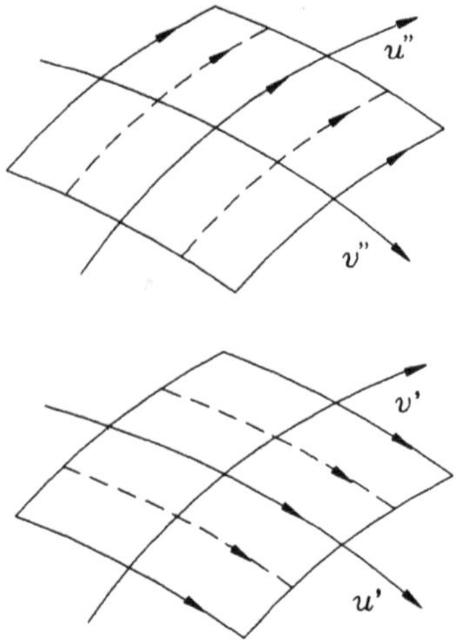

Figure 3.9 Two overlapping generalised quadrilaterals with u and v co-ordinates interchanged

element we have a simple interchange of the variables, for a triangular element we have to define, in general, three different local co-ordinate systems. This obviously makes triangular elements less convenient than quadrilateral ones.

Consider finally the case of a degenerate quadrilateral of the form in Figure 3.7, i.e. a generalised tubular surface, from this point of view. If the tube is electrically thin, it is possible to assume that the surface-current-density vector has no circumferential component (the v component in Figure 3.7) at all, but

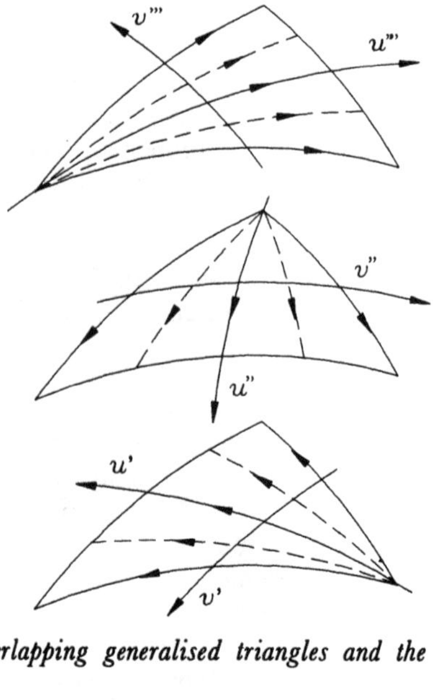

Figure 3.10 Three overlapping generalised triangles and the corresponding co-ordinate systems

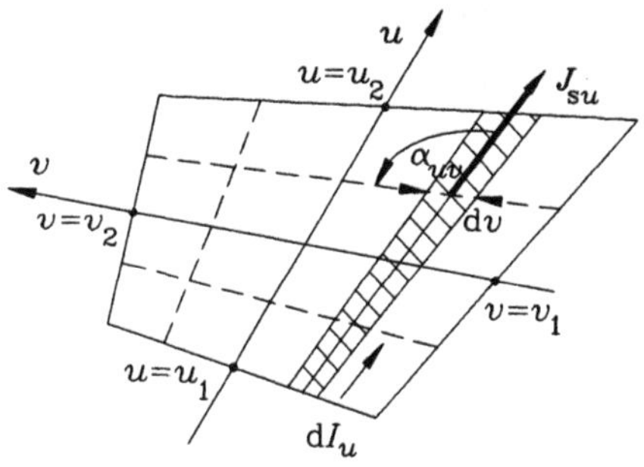

Figure 3.11 *Definitions of some quantities relating to currents over generalised quadrilaterals*

only the u component, which makes the approximation much simpler. However, such an approximation is not admissible for tubular surfaces of arbitrary size, in which case the procedure outlined for a quadrilateral (two overlapping tubular surfaces with interchanged co-ordinate systems) should be applied.

Thus, the components of the surface-current-density vector can in all cases be regarded as (for example) the u components on curved quadrilaterals (normal or degenerate) which overlap or are interconnected. Therefore in the theoretical analysis which follows only such quadrilaterals, carrying only the u component of the current, need be considered. The surface-current-density vector over such a curved quadrilateral can be represented as

$$\boldsymbol{J}_s = J_{su}(u, v)\boldsymbol{i}_u(u, v) \tag{3.52}$$

To determine the surface-charge density associated with this surface-current density, consider the current in the elemental strip of local width dv, as in Figure 3.11:

$$\mathrm{d}I_u = J_{su}\,\mathrm{d}l_v \sin\alpha_{uv} = J_{su}\,e_v \sin\alpha_{uv}\,\mathrm{d}v \tag{3.53}$$

α_{uv} representing the angle between the local u and v co-ordinate lines. To this line current corresponds the charge per unit length

$$\mathrm{d}Q'_u = \rho_{su}\,\mathrm{d}l_v \sin\alpha_{uv} = \rho_{su}\,e_v \sin\alpha_{uv}\,\mathrm{d}v \tag{3.54}$$

Substituting the expressions in eqns. 3.53 and 3.54 into the continuity equation valid for line currents

$$\frac{\mathrm{d}}{\mathrm{d}l_u}(\mathrm{d}I_u) = -j\omega(\mathrm{d}Q'_u) \tag{3.55}$$

we obtain the expression for the surface-charge density ρ_s corresponding to the surface-current-density vector $\boldsymbol{J}_s$ given in eqn. 3.52:

$$\rho_s = \rho_{su} = \frac{j}{\omega}\frac{1}{e_u e_v \sin\alpha_{uv}}\frac{\mathrm{d}}{\mathrm{d}u}(J_{su}\,e_v \sin\alpha_{uv}) = \frac{j}{\omega}\,\mathrm{div}_s\boldsymbol{J}_s \tag{3.56}$$

For further analysis it will be necessary to know the current intensity per unit of the v co-ordinate. From eqns. 3.53 it can be expressed as

$$\frac{dI_u}{dv} = J_{su}\, e_v \sin\alpha_{uv} \tag{3.57}$$

Of course, if v is a length co-ordinate dI_u/dv has the dimension of surface-current density.

3.3.3 Basis functions automatically satisfying the continuity equation along interconnections and edges of generalised quadrilaterals

Consider the interconnection of two quadrilaterals sketched in Figure 3.12 and consider, for the moment, the current over one of the quadrilaterals only.

The continuity equation at the interconnection is expressed in terms of the u component of the current, the v component being tangential to the line defining the interconnection. However, since in general the u and v co-ordinate lines at the interconnection are not mutually perpendicular, to formulate the continuity equation we need only the projection of the surface-current-density vector onto the normal to the local v co-ordinate line defining the interconnection and lying in the local tangential plane to the quadrilateral surface. Thus, with reference to Figure 3.12 and bearing in mind eqn. 3.57, the normal component of the surface-current-density vector along the line $u = u_1$ can be expressed as

$$J_{su}(u_1, v)_{norm} = J_{su}(u_1, v)\sin\alpha_{uv}(u_1, v) = \frac{1}{e_v(u_1, v)} \left.\frac{dI_u}{dv}\right|_{u=u_1} \tag{3.58}$$

If needed, an analogous expression can be written for the normal component of the surface-current-density vector along the line $u = u_2$.

Note that along the interconnection the Lamé coefficients $e_v(u_1, v)$ for the two quadrilaterals are of the same form. This follows from the definition of the Lamé coefficients $dl_v = e_v(u_1, v)\, dv$, dl_v being the same length element (along the interconnection) for both quadrilaterals. If, in addition, dv along the interconnection is the same for the two quadrilaterals, the Lamé coefficients become identical. We shall assume henceforth that, for any strict intercon-

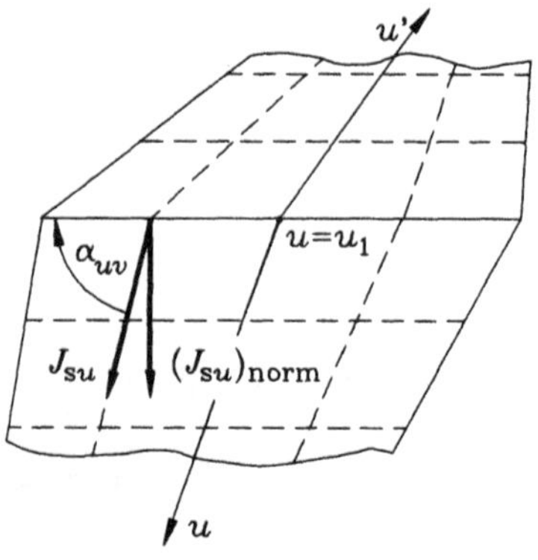

Figure 3.12 Interconnection of two generalised quadrilaterals

nection, this always is the case. From eqn. 3.58 it then follows that the currents dI_u/dv for the two quadrilaterals at the interconnection must be constructed from the basis functions of the same form, since otherwise the continuity equation cannot be satisfied. As a result of this conclusion, it appears to be extremely convenient to express the surface-current density over generalised quadrilaterals in terms of the current dI_u/dv, so that the continuity equation can be formulated with ease. [Note also that the angles $\alpha_{uv}(u_1, v)$ for surface-current-density vectors at the interconnection are generally different for the two quadrilaterals.] For this reason, the following initial approximation for the current distribution has been adopted in this book (see eqn. 3.53):

$$J_{su}(u, v) = \frac{1}{e_v \sin\alpha_{uv}} \frac{dI_u}{dv} \tag{3.59}$$

$$\frac{dI_u}{dv} = \sum_{j=1}^{n_v} \left\{ \sum_{i=1}^{n_u} a_{ij} f_i(u) \right\} h_j(v) \tag{3.60}$$

where n_u and n_v are the orders of the approximations along the u and v co-ordinates, a_{ij} are unknown coefficients to be determined, and $f_i(u)$ and $h_j(v)$ are known arbitrary basis functions.

We now wish to construct from the expression in eqns. 3.59 and 3.60 basis functions which satisfy the continuity equation along the interconnection of two quadrilaterals. To that aim we use a procedure similar to that in References 64 and 67. First, let us write the expressions for dI_u/dv at points along the lines $u = u_1$ and $u = u_2$ in the form

$$\left.\frac{dI_u}{dv}\right|_{u=u_1} = \sum_{j=1}^{n_v} b_{1j} h_j(v) \qquad \left.\frac{dI_u}{dv}\right|_{u=u_2} = \sum_{j=1}^{n_v} b_{2j} h_j(v) \tag{3.61}$$

where b_{1j} and b_{2j} are unknown coefficients defined by the equations

$$b_{1j} = \sum_{i=1}^{n_u} a_{ij} f_i(u_1) \qquad b_{2j} = \sum_{i=1}^{n_u} a_{ij} f_i(u_2) \qquad j = 1, \ldots, n_v \tag{3.62}$$

Starting from this equation pair, for any j it is possible to express two unknown coefficients, for example a_{1j} and a_{2j}, in terms of the others, a_{ij}, $i = 3, \ldots, n_u$, and of the coefficients b_{1j} and b_{2j}. Substituting the expressions for a_{1j} and a_{2j} thus obtained into eqn. 3.60, and this equation into eqn. 3.59, the u component $J_{su} = J_{su}(u, v)$ of the surface-current-density vector can be expressed as

$$J_{su} = \frac{1}{e_v \sin\alpha_{uv}} \sum_{j=1}^{n_v} \left\{ b_{1j} g_1(u) + b_{2j} g_2(u) + \sum_{i=3}^{n_u} a_{ij} g_i(u) \right\} h_j(v) \tag{3.63}$$

where the expressions for $g_i(u)$, $i = 1, 2, \ldots, n_u$, are the same as for generalised wires, eqns. 3.6, 3.7, 3.9 and 3.10, which we repeat here for completeness:

$$g_1(u) = \frac{1}{Q_{12}} \{f_2(u_2) f_1(u) - f_1(u_2) f_2(u)\} \tag{3.64}$$

$$g_2(u) = \frac{1}{Q_{21}} \{f_2(u_1)f_1(u) - f_1(u_1)f_2(u)\} \tag{3.65}$$

$$g_i(u) = f_i(u) + \frac{1}{Q_{21}} \{Q_{i2}f_1(u) - Q_{i1}f_2(u)\} \qquad i = 3, \ldots, n \tag{3.66}$$

where

$$Q_{ij} = f_i(u_1)f_j(u_2) - f_i(u_2)f_j(u_1) \qquad i = 1, 2, \ldots, n \qquad j = 1, 2 \tag{3.67}$$

Note that along the line $u = u_1$ only $g_1(u)$ is nonzero $[g_1(u_1) = 1]$, and that along the line $u = u_2$ only $g_2(u)$ is nonzero $[g_2(u_2) = 1]$.

The expansion in eqn. 3.63 partially fulfills the desired requirements. It can be used directly to satisfy the current-continuity conditions along free sides of generalised quadrilaterals and at the same time to reduce the number of unknowns. For example, if the side $u = u_1$ of the quadrilateral is free (not connected to any other quadrilateral), in the expansion (eqn. 3.63) we set simply $b_{1j} = 0$, $j = 1, 2, \ldots, n_v$. If instead the side $u = u_2$ is free, we set $b_{2j} = 0$, $j = 1, 2, \ldots, n_v$. If both sides are free, we set $b_{1j} = b_{2j} = 0$, $j = 1, 2, \ldots, n_v$.

As with generalised wires, in this case the chief usefulness of the expansion in eqn. 3.63 is again not in simple formulation of the current-continuity equation at the free sides of quadrilaterals. Rather, it is much more useful for formulation of the expansion which automatically satisfies the current-continuity equation along a common side of any number of interconnected generalised quadrilaterals.

To derive such an expansion, consider m quadrilaterals interconnected along the sides $u^{(k)} = u_1^{(k)}$, $k = 1, 2, \ldots, m$. The co-ordinate u_1 differs from one quadrilateral to another, but for simplicity we shall henceforth omit the superscript k in $u_1^{(k)}$. (It is always possible to adopt local-co-ordinate systems for the quadrilaterals such that, for all the quadrilaterals, the common side coincides with the line $u = u_1$.) Assume that we have found current expansions for the u component of the surface-current-density vector over individual quadrilaterals in the form given in eqn. 3.63,

$$J_{su}^{(k)} = \frac{1}{e_v^{(k)} \sin \alpha_{uv}^{(k)}} \sum_{j=1}^{n_v} \left\{ b_{1j}^{(k)} g_1^{(k)}(u) + b_{2j}^{(k)} g_2^{(k)}(u) + \sum_{i=3}^{n_u^{(k)}} a_{ij}^{(k)} g_i^{(k)}(u) \right\} h_j(v) \qquad k = 1, 2, \ldots, m \tag{3.68}$$

Note that for all quadrilaterals sharing the common side $u = u_1$ the functions $h_j(v)$ must be the same, since otherwise the current-continuity equation cannot be satisfied. However, the functions $g_i^{(k)}(u)$, $i = 1, 2, \ldots, n_u^{(k)}$, as well as $n_u^{(k)}$, $k = 1, 2, \ldots, m$, are obviously arbitrary as far as the common side $u = u_1$ of the quadrilaterals is concerned.

For n_v, it must be the same for the expansions over the m quadrilaterals along the side they share, but it is possible to adopt different values of n_v for the three terms in the curly brackets of eqn. 3.68. This will be discussed later in this Section.

The local-currrent-continuity condition now has the form

$$\sum_{k=1}^{m} (J_{su})_{norm}^{(k)}(u_1, v) = 0 \tag{3.69}$$

where the subscript *norm* denotes the component of J_{su} over the quadrilaterals along the line $u = u_1$ and normal to it. Substituting the expression in eqn. 3.58 into eqn. 3.69, we obtain

$$\sum_{k=1}^{m} \frac{1}{e_v^{(k)}(u_1, v)} \left. \frac{dI_u^{(k)}}{dv} \right|_{u=u_1} = 0 \tag{3.70}$$

Bearing in mind the expression in eqn. 3.61, and that $e_v^{(k)}(u_1, v)$ is the same for all the quadrilaterals sharing the common side (i.e. for all $k = 1, 2, \ldots, m$), the continuity equation is reduced to the system of equations

$$\sum_{k=1}^{m} b_{1j}^{(k)} = 0 \qquad j = 1, 2, \ldots, n_v \tag{3.71}$$

Starting from the above equation for all values of j, it is possible to express one of the unknowns $b_{1j}^{(k)}$, $k = 1, \ldots, m$, e.g. $b_{1j}^{(1)}$, in terms of the others. By substituting the expression thus obtained for $b_{1j}^{(1)}$ into eqn. 3.68 for $k = 1$, an expansion $(J_{su})^{(1)}$ for the quadrilateral labelled 1 is obtained which automatically satisfies the current-continuity equation along the interconnection:

$$(J_{su})^{(1)} = \frac{1}{e_v^{(1)} \sin \alpha_{uv}^{(1)}} \sum_{j=1}^{n_v} \left[-\left\{ \sum_{k=2}^{m} b_{1j}^{(k)} \right\} g_1^{(1)}(u) + b_{2j}^{(1)} g_2^{(1)}(u) + \sum_{i=3}^{n_u^{(1)}} a_{ij} g_i^{(1)}(u) \right] h_j(v) \tag{3.72}$$

The expansions for other generalised quadrilaterals sharing the common side, $k = 2, \ldots, m$, remain as in eqn. 3.68. We have thus established the required expansions for the surface current over the *m* quadrilaterals sharing a common side which automatically satisfy the current-continuity equation along that side.

Along the opposite sides of the quadrilaterals with respect to the intersection (i.e. the sides defined by $u = u_2$), only the second term on the right-hand side in eqns. 3.68 and 3.72 is nonzero. Therefore, if that other side of a quadrilateral is free, we omit this term. If it is not, i.e. if it is shared by other quadrilaterals, it must be present. We combine it with the corresponding terms in the expansions over these quadrilaterals, as outlined above, to obtain the expansions for quadrilaterals sharing that other side which also satisfy the continuity equation along it. Only if the quadrilateral has both $u = u_1$ and $u = u_2$ sides free (i.e. not shared with any other quadrilateral) does one type of expansion suffice, in the form of the sum in eqn. 3.63.

Note that the first term on the right-hand side in eqn. 3.72, when compared with that in eqn. 3.68, is decomposed into $(m-1)$ terms. We can combine each of the terms under the first summation sign on the right-hand side of eqn. 3.72 with the first term in eqn. 3.68, $k = 2, 3, \ldots, m$, and thus obtain the following expansions extending over the generalised quadrilateral labelled 1 and generalised quadrilaterals labelled 2, 3, . . . , m, which are continuous along the whole length of the common side, $u = u_1$, of the m quadrilaterals:

$$(J_{su})_d^{(k)} = \sum_{j=1}^{n_v} b_{1j}^{(k)} \left\{ \begin{array}{ll} \dfrac{-g_1^{(1)}(u)}{e_v^{(1)} \sin \alpha_{uv}^{(1)}} & u_1^{(1)} \leqslant u \leqslant u_2^{(1)} \\ \dfrac{g_1^{(k)}(u)}{e_v^{(k)} \sin \alpha_{uv}^{(k)}} & u_1^{(k)} \leqslant u \leqslant u_2^{(k)} \end{array} \right\} h_j(v)$$

$$k = 2, 3, \ldots, m \qquad (3.73)$$

An analogous treatment is possible for the other sides of the quadrilaterals.

We see that, as for current expansions along generalised wires, the functions $g_1 h_j$, $g_2 h_j$ and $g_i h_j$, $i = 3, \ldots, n_u$, for all the generalised quadrilaterals which approximate a body considered can be categorised in two expansion groups. To the first group belong the expansions consisting of functions $g_i h_j$, $i = 3, \ldots, n_u$, which are defined over one generalised quadrilateral, and which for any quadrilateral are of the form

$$(J_{su})_s = \frac{1}{e_v \sin \alpha_{uv}} \sum_{j=1}^{n_v} \left\{ \sum_{i=3}^{n_u} a_{ij} g_i(u) \right\} h_j(v) \qquad (3.74)$$

These expansions are zero along the lines $u = u_1$ and $u = u_2$, thus automatically satisfying the continuity equation along free edges of a surface element and having no influence on the continuity of current at interconnections of two surface elements.

The second group make the expansions consisting of the functions $g_1 h_j$ and $g_2 h_j$, which influence the continuity of current along only one side ($u = u_1$ or $u = u_2$) of two interconnected quadrilaterals. As has been pointed out, any interconnection can be treated as an interconnection of two quadrilaterals interconnected along the side $u = u_1$. In that case all the terms of the form $b_{1j}^{(k)} g_1^{(k)} h_j$, $k = 1, \ldots, m$, where m is the number of the surfaces at the interconnection, can be grouped in an expansion of the form given in eqn. 3.73. This expansion is zero along the lines $u^{(k)} = u_2^{(k)}$, so that it does not enter the continuity equation for other interconnections. For the lines $u^{(k)} = u_2^{(k)}$ representing free quadrilateral sides, the above expansion automatically satisfies the continuity equation along it.

In this manner, starting from arbitrary functional series in eqn. 3.60, we have obtained an entire-domain expansion in eqns. 3.73 and 3.74 which automatically satisfies the continuity equation along the surface-element interconnections and free edges. Since the basis functions in the expansion of eqn. 3.74 are defined over a single surface, we term them 'singletons'. In analogy, the basis functions in the expansion of eqn. 3.73 being defined over pairs of two interconnected surface elements, we term them 'doublets'. Finally, note that for the lowest degree of approximation ($n_v = 1$ and $n_u = 2$ in eqn. 3.60) the singletons are lost, and the doublets have the form of the classical subdomain functions.

It will be demonstrated that, by appropriate choice of the functions $f_i(u)$ and $h_j(v)$, it is possible to obtain very simple singletons and doublets.

3.3.4 Polynomial entire-domain current expansions over generalised quadrilaterals

In an analogous way to the definition of entire-domain current expansions for generalised wires in Section 3.2.3, in exactly the same manner we define the type of current expansion along one of the co-ordinates, u or v, of a generalised quadrilateral. Thus, for example, we can have entire-domain expansion along one co-ordinate, and subdomain expansion along the other. Usually, however, we have the same type of expansion in both co-ordinates. Therefore, for example, 'entire-domain expansion' for a generalised quadrilateral means that it is of the entire-domain type in both co-ordinates.

Among many possibilities for the functions f_i and h_j in eqn. 3.60 that can be chosen, probably the most convenient is that f_i and h_j be adopted as power functions. There are several reasons for this particular choice. First, the polynomials are very flexible: with only a few terms, very diverse function shapes can be approximated fairly accurately. Secondly, polynomials are computed quickly; this is also important, because they have to be evaluated many times in the analysis process. Thirdly, in some cases the integrals with powers of higher degrees can be obtained, using a recursive formula, from those of lower degree.

As explained in Section 2, we consider all surface currents as the u components in the co-ordinate systems of generalised quadrilaterals which are interconnected or overlap. As mentioned, we approximate the surface-current density by an entire-domain polynomial expansion. The basis functions of this type automatically satisfying the continuity equation along the generalised quadrilateral interconnections and free edges are obtained from the general expression in Section 3.3.3. By setting $f_i(u) = u^{i-1}$, $i = 1, \ldots, n_u$, $-1 \leqslant u \leqslant 1$, and $h_j(v) = v^{j-1}$, $j = 1, \ldots, n_v$, $-1 \leqslant v \leqslant 1$, in eqn. 3.63, a very simple current expansion over a generalised quadrilateral is obtained:

$$J_{su} = \frac{1}{e_v \sin \alpha_{uv}} \sum_{j=1}^{n_v} \left\{ b_{1j} \frac{1-u}{2} + b_{2j} \frac{1+u}{2} + \sum_{\substack{i=3 \\ (2)}}^{n_u} a_{ij}(u^{i-1} - 1) + \sum_{\substack{i=4 \\ (2)}}^{n_u} a_{ij}(u^{i-1} - u) \right\} v^{j-1} \tag{3.75}$$

The singleton expansion for a generalised quadrilateral in this case is thus

$$(\mathcal{J}_{su})_s = \frac{1}{e_v \sin \alpha_{uv}} \sum_{j=1}^{n_v} \sum_{i=3}^{n_u} a_{ij} \left\{ \begin{matrix} u^{i-1} - 1, & i\,\text{odd} \\ u^{i-1} - u, & i\,\text{even} \end{matrix} \right\} v^{j-1} \tag{3.76}$$

The doublets, for example those defined on the kth and the first quadrilateral

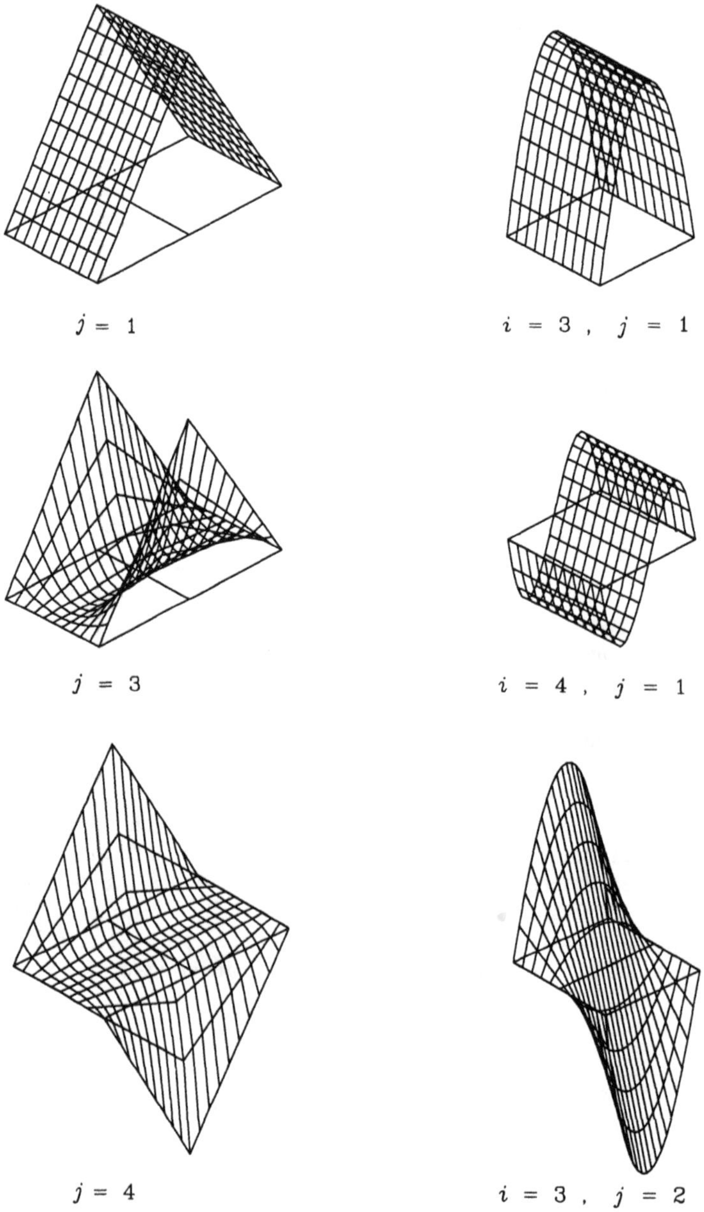

Figure 3.13 First few terms of the series in eqns. 3.76 and 3.77

having a common side, with $u^{(1)} = -1$ and $u^{(k)} = -1$ along the interconnection of m quadrilaterals, are

$$(J_{su})_d^{(k)} = \sum_{j=1}^{n_v} b_{1j}^{(k)} \left\{ \begin{array}{ll} \dfrac{u^{(1)} - 1}{e_v^{(1)} \sin \alpha_{uv}^{(1)}} & -1 \leqslant u^{(1)} \leqslant 1 \\ \dfrac{1 - u^{(k)}}{e_v^{(k)} \sin \alpha_{uv}^{(k)}} & -1 \leqslant u^{(k)} \leqslant 1 \end{array} \right\} v^{j-1},$$

$$-1 \leqslant v \leqslant 1 \qquad k = 1, 2, \ldots, m \qquad (3.77)$$

The first few terms of the series in eqns. 3.76 and 3.77 are sketched in Figure 3.13. Note that the series in eqns. 3.76 and 3.77 can easily take into account possible symmetry or antisymmetry of a structure, by simply omitting the odd (even) terms.

By numerical experiment it was found that polynomial expansions result in acceptable current distributions over large surfaces with only about ten unknowns per wavelength squared for both current components (e.g. Reference 67).

3.3.5 Subdomain current expansions

We have defined subdomain current expansions as those obtained with the lowest order of approximation of current ($n_v = 1$ and $n_u = 2$ in the expression of eqn. 3.60) over all quadrilaterals which guarantees current continuity from one quadrilateral to the next. The most important subdomain approximations over quadrilaterals are obtained as generalisations of subdomain approximations along wires. The approximation along the transversal co-ordinate (i.e. the v co-ordinate) is usually a piecewise-constant approximation, and the approximation along the longitudinal co-ordinate (i.e. the u co-ordinate) is either a piecewise-linear or a piecewise-sine approximation. For such approximations it is possible to construct basis functions which automatically satisfy the continuity equation at free sides and interconnections of generalised quadrilaterals. Thus the so-called rooftop functions are obtained for piecewise-linear approximation, and the so-called surface sine doublets are obtained for piecewise-sine approximation.

The rooftop functions represent a special case of polynomial expansions given in Section 3.3.4, and which is completely defined by eqn. 3.77. It is well known that to obtain satisfactory results for current distribution, such an expansion requires about 100 unknowns per wavelength squared for each current component.

The surface sine doublet can be considered as a special case of a combined polynomial/trigonometric expansion for quadrilaterals analogous to that for wires described in Section 3.2.3. For m quadrilaterals sharing a common side, with the starting side of all quadrilaterals at the interconnection (i.e. at $u_1^{(k)} = -1$, $k = 1, 2, \ldots, m$), the $(m-1)$ surface sine doublets are defined as

$$(J_{su})_d^{(k)} = b_{11}^{(k)} \left\{ \begin{array}{ll} \dfrac{\sin[kl^{(1)}\{u^{(1)} - 1\}]}{e_v^{(1)} \sin \alpha_{uv}^{(1)} \sin\{2kl^{(1)}\}} & -1 \leqslant u^{(1)} \leqslant 1 \\ \dfrac{\sin[kl^{(k)}\{1 - u^{(k)}\}]}{e_v^{(k)} \sin \alpha_{uv}^{(k)} \sin\{2kl^{(k)}\}} & -1 \leqslant u^{(k)} \leqslant 1 \end{array} \right\}$$

$$k = 2, 3, \ldots, m \qquad (3.78)$$

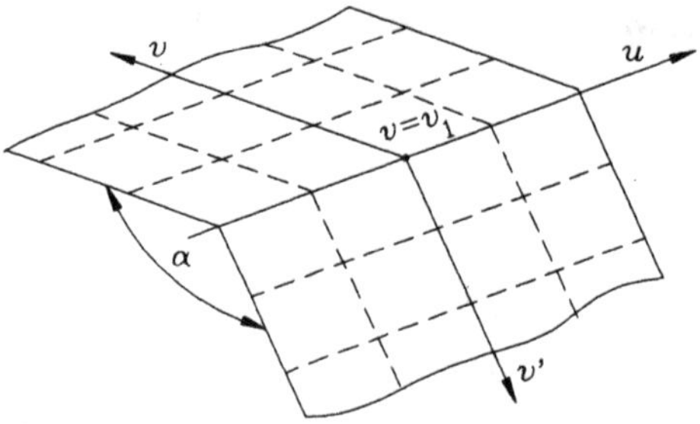

Figure 3.14 A perfectly conducting wedge

where $u^{(k)}$ is the local nondimensional co-ordinate, and the length of the side of the kth quadrilateral is $2l^{(k)}$. To obtain the current distribution to acceptable accuracy this expansion is known to require about 30 unknowns per wavelength squared for both current components.

It is obvious that the number of unknowns needed for any subdomain approximation of current over a quadrilateral is on average much higher than that required by an entire-domain approximation. Although subdomain approximation has been used quite frequently, the authors believe that, because of this major disadvantage when compared with the entire-domain approximation, it does not warrant any further attention in this book.

3.3.6 *Quasistatic treatment of edges and corners*

The distribution of electromagnetic field in electrically small domains about edges and corners is very nearly quasistatic. The overall accuracy of the results and their stability may be improved somewhat by taking this fact into account.

As an example, consider the edge of a wedge (Figure 3.14). The distribution of the surface charge in the vicinity of the edge can be expressed as [86]

$$\rho_s(u, v) = a(u)(v - v_1)^b \qquad b = \pi/(2\pi - \alpha) \tag{3.79}$$

where $a(u)$ is a well behaved function of u, adopted to be the local co-ordinate axis parallel to the edge, v is the local co-ordinate axis perpendicular to the edge, v_1 is the v co-ordinate of the edge and α is the angle between the two planes forming the wedge. Note that b takes values from $b = -1$ to $b = 0$ when α increases from $\alpha = 0$ to $\alpha = \pi$. Therefore, if $\alpha < \pi$, the surface-charge density at the edge ($v = v_1$) is infinite, i.e. the charge distribution in the vicinity of the edge is quasisingular.

From the continuity equation for surface currents it follows that the surface-current component parallel to the edge exhibits the same quasisingular behaviour as the surface charge in the vicinity of the edge. This property can be taken into account if the functions $h_j(v)$, $j = 1, 2, \ldots, n_v$, in the expansion in eqn. 3.68 are adopted in the form suggested by eqn. 3.79. Such basis functions have been used to improve the convergence of the results (e.g. Reference 87). Although this does have certain advantages, in the approach adopted in this book this is not indispensable, since for sufficiently small surface elements and with a sufficiently high degree for the current approximation, any distribution of currents and charges can be approximated with reasonable accuracy.

The corner of a plate and the vertex of a body approximated by plates can be considered as special cases of the generalised cones shown in Figure 3.5 in Section 3.2.5. From that Section, the current flowing into (or out of) the tip of the generalised cone should be zero and the corresponding charge distribution per unit length of the cone generatrix should also be zero. It can be shown that these constraints are automatically satisfied for expansions which satisfy the continuity equation along edges and interconnections of nondegenerate generalised quadrilaterals. The explanations in this Section indicate that, in the approach adopted in this book, no additional basis functions based on quasistatic analysis are indispensable in treating edges and corners of a metallic structure.

3.3.7 Choice of basis functions over generalised triangles

The current expansions in eqns. 3.73 and 3.74 are also valid if the approximation of a structure is performed by triangles and the surface current is decomposed as in Figure 3.10. However, in this case the current expansions have to satisfy additional constraints. Consider a triangle with the u co-ordinate lines intersecting at the point $u = u_1$. By definition, any elemental length can be expressed as $dl_v = e_v dv$. The Lamé coefficient $e_v(u, v)$ therefore tends to zero when u tends to u_1. Bearing in mind eqn. 3.59, and noting that the u component of the surface-current-density vector at that point can only be zero, as explained, dI_u/dv must tend to zero as $(u - u_1)^2$. Therefore the additional constraints can be written in the form

$$\left.\frac{dI_u}{dv}\right|_{u=u_1} = 0 \tag{3.80}$$

$$\left.\frac{d}{du}\left(\frac{dI_u}{dv}\right)\right|_{u=u_1} = 0 \tag{3.81}$$

It is obvious that the first constraint can be fulfilled by simply omitting the doublet expansion at the point $u = u_1$. Substituting eqn. 3.60 into eqn. 3.81, the second constraint can be expressed as

$$\left.\frac{df_i(u)}{du}\right|_{u=u_1} = 0 \tag{3.82}$$

This constraint can easily be satisfied by proper choice of the initial basis functions f_i, $i = 1, 2, \ldots, n_u$.

Very simple current expansions are obtained if the polynomial approximation is adopted and if the starting and end values of the u and v co-ordinates are adopted to be $u_1 = 0$, $u_2 = 1$, $v_1 = -1$, and $v_2 = 1$. In that case, the initial basis functions can be written in the form $f_1(u) = 1, f_i(u) = u^i$, $0 \leqslant u \leqslant 1$, $i = 2, \ldots, n_u$, and $h_j(v) = v^{j-1}$, $j = 1, \ldots, n_v$, $-1 \leqslant v \leqslant 1$. Note that these initial basis functions are the same as those used in the polynomial expansion over generalised

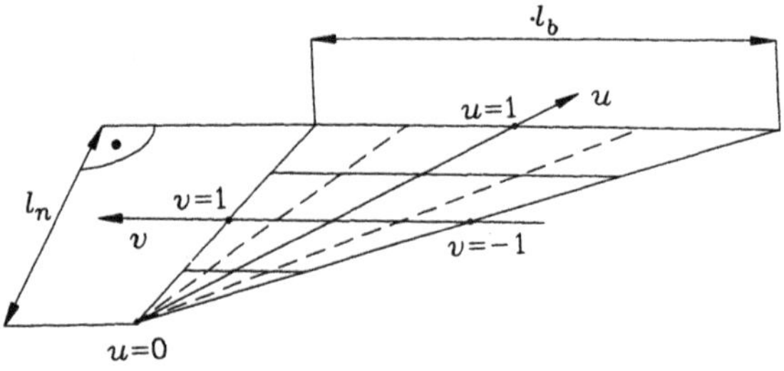

Figure 3.15 A flat triangular element

quadrilaterals (see eqn. 3.75), except that $u_1 = 0$ and that the term obtained for $i = 2$ is omitted. Starting from these basis functions the following expansion is obtained:

$$\mathcal{J}_{su} = \frac{1}{e_v \sin \alpha_{uv}} \sum_{j=1}^{n_v} \left\{ b_{2j} u^2 + \sum_{i=3}^{n_u} a_{ij}(u^i - u^2) \right\} v^{j-1} \tag{3.83}$$

For flat triangles this expansion can be simplified further. Consider a flat triangle with the starting and end u and v co-ordinates as adopted above (Figure 3.15). The parametric equation of the triangle can be written in the form (see eqn. 2.28)

$$\boldsymbol{r}(u, v) = \boldsymbol{r}_u u + \boldsymbol{r}_{uv} uv \qquad 0 \leqslant u \leqslant 1 \qquad -1 \leqslant v \leqslant 1 \tag{3.84}$$

where it is assumed that $\boldsymbol{r}_c = 0$. Starting from this equation, the Lamé coefficient e_v in eqn. 3.83 can be expressed as $e_v = |\boldsymbol{r}_{uv}| u = l_b u/2$, where l_b is the length of the triangle basis. In addition, the vector $\boldsymbol{r}(u, v)$ can be expressed as $\boldsymbol{r} = (ul_n/\sin\alpha_{uv})\boldsymbol{i}_u$, where l_n is the length of the triangle height. Remembering that the area of the triangle is $S = l_b l_n/2$, the expansion in eqn. 3.83 can be written in the vector form

$$\boldsymbol{J}_{su} = \frac{\boldsymbol{r}(u, v)}{S} \sum_{j=1}^{n_v} \left\{ b_{2j} + \sum_{i=3}^{n_u} a_{ij}(u^{i-2} - 1) \right\} v^{j-1} \tag{3.85}$$

Specifically, for $n_u = 2$ and $n_v = 1$, the subdomain approximation is obtained. Just as for generalised quadrilaterals, the subdomain approximation can be expressed in the form of doublets. The doublets defined on the kth and the first triangle sharing a common side, with $u^{(1)} = 1$ and $u^{(k)} = 1$ along the interconnection of m triangles, can be written as

$$(\mathcal{J}_{su})_d^{(k)} = b_{21}^{(k)} \left\{ \begin{array}{ll} \dfrac{\boldsymbol{r}^{(1)}(u, v)}{S^{(1)}} & 0 \leqslant u \leqslant 1 \\ -\dfrac{\boldsymbol{r}^{(k)}(u, v)}{S^{(k)}} & 0 \leqslant u \leqslant 1 \end{array} \right\} \qquad k = 1, 2, \ldots, m \tag{3.86}$$

These doublets represent triangular doublets [49], known to be of approximately the same efficiency as the rooftop functions.

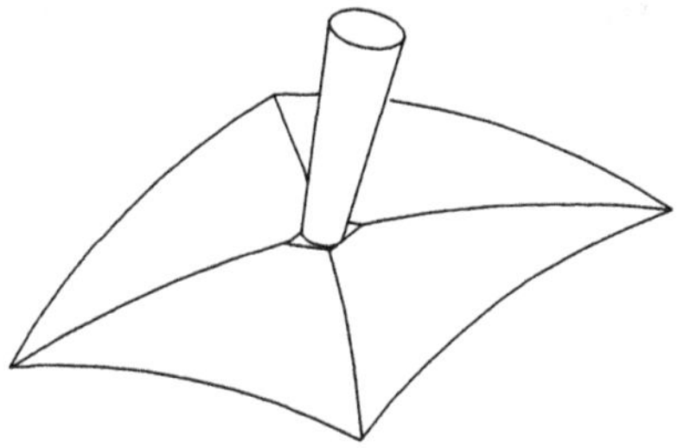

Figure 3.16 *Interconnection of a right-truncated cone and a plate partitioned into optimum generalised quadrilaterals*

3.4 Current expansions across junctions of incompatible elements

3.4.1 Introduction

All interconnections encountered in structures approximated by plates and (generalised) wires belong to three principal classes. To the first class belong those which satisfy the two conditions:

- all interconnected elements have identical common sides
- all interconnected elements are such that the natural order of approximation along the interconnection is the same for all of them.

For example, the junction of a cylindrical wire of any radius and its conical end, or the interconnection of two or more plates of approximately the same size sharing a common side, belong to this class.

To the second class belong those interconnections which satisfy the first condition but not the second. For example, the interconnection of a small quadrilateral, or of a narrow strip along its narrow side, with a quadrilateral that has a very long side opposite to that at the interconnection belongs to this class.

Finally, we are forced frequently to model an interconnection in such a manner that the first condition is not satisfied, although the second condition may be. Such cases belong to the third class. One example is the interconnection of two or more wires of arbitrary radius meeting at an arbitrary angle. We always model wire segments by right circular cylinders; therefore at the interconnection they share at the most only a few common points. Another important example is the interconnection at any angle of a (generalised) wire and a plate. We chose to model the wire by a right truncated cone. To approximate the junction optimally, the plate must be partitioned into convenient curvilinear quadrilaterals, as shown in Figure 3.16. It is evident, however, that it is never possible to satisfy the first condition.

The treatment of the first class of junctions has been explained. It poses no conceptual or computational problems. The treatment of the other two classes of junctions, however, requires additional consideration. This Section is aimed at considering one type of interconnection of the second class and two of the third class, precisely those mentioned above as examples. Following similar lines of reasoning it is possible to treat other specific junctions not belonging to the first class.

3.4.2 *Junctions of small and large quadrilaterals*

It may be necessary, or at least advantageous, to approximate the surface currents over two generalised quadrilaterals sharing a common side by expansions of different degrees n_v in the v co-ordinate. For example, for the approximation of geometry of a structure we may need an electrically relatively small element as one quadrilateral, sharing a side with a quadrilateral which spreads out and has a much longer opposite side. The value of n_v needed for the small quadrilateral is low, but along the interconnection that same n_v must have the expansion for the other quadrilateral. However, while the small value of n_v may (and usually will) be appropriate for the entire small quadrilateral, it may not be for the entire surface of the large one. We can, of course, adopt the higher n_v of the two (needed for the large quadrilateral) for the small quadrilateral as well, but this would needlessly increase the number of unknowns. Instead, we can adopt different degrees n_v in the first two sums on the right-hand side of eqn. 3.68 in the expansions for the small and large quadrilateral, as follows.

Considering again the general case of m quadrilaterals meeting at the interconnection, we conclude that n_v must be the same for those parts of the expansions for the m quadrilaterals that are nonzero along the side they share. However, $g_1^{(k)}(u_1)$ is the only nonzero function along that side, while it is zero along the opposite quadrilateral side ($u = u_2$). Along the side $u = u_2$, only $g_2^{(k)}(u)$ is nonzero. Therefore we can adopt different values of n_v for these two terms, e.g. n_{v1} and n_{v2}. In fact, if desired, we can also adopt a different value of n_v, e.g. n_{v3}, for the third term. Thus we can rewrite eqn. 3.68 in the form

$$J_{su}^{(k)} = \frac{1}{e_v^{(k)} \sin \alpha_{uv}^{(k)}} \left\{ g_1^{(k)}(u) \sum_{j=1}^{n_{v1}^{(k)}} b_{1j}^{(k)} h_j(v) + g_2^{(k)}(u) \sum_{j=1}^{n_{v2}^{(k)}} b_{2j}^{(k)} h_j(v) + \sum_{j=1}^{n_{v3}^{(k)}} \sum_{i=3}^{n_u^{(k)}} a_{ij}^{(k)} g_i^{(k)}(u) h_j(v) \right\} \qquad k = 1, 2, \ldots, m \qquad (3.87)$$

Assume that the lowest n_v needed by a quadrilateral along the interconnection is n_{vmin}. We then adopt $n_{v1}^{(k)} = n_{vmin}$, $k = 1, 2, \ldots, m$. As far as the side $u = u_1$ is concerned, the values of $n_{v2}^{(k)}$ and $n_{v3}^{(k)}$, $k = 1, 2, \ldots, m$, can be adopted at will.

Note that by this procedure we have not, in fact, reduced the order of approximation anywhere except along the common side. Over the rest of the quadrilaterals the order of approximation is determined by all the terms, i.e. by the other doublet and all the singletons.

3.4.3 *Wire-to-plate junctions*

It was explained in Chapter 2 that the treatment of wire-to-plate junctions is based on a general localised junction model. According to this model, any localised junction can be considered to consist of the ends of generalised wires and adjacent sides of generalised quadrilaterals (plates) situated in an electrically small junction domain. Since any generalised wire can be treated as a generalised quadrilateral, consider a localised junction in the form of an interconnection of m quadrilaterals, with sides $u^{(k)} = u_1^{(k)}$ situated in the junction domain. Bearing in mind that the junction domain is electrically small, the current expansion over all the quadrilaterals can be written as in eqn. 3.87, but with $n_{v1}^{(k)} = 1$, i.e. in the form

$$\mathcal{J}_{su}^{(k)} = \frac{1}{e_v^{(k)} \sin \alpha_{uv}^{(k)}} \left\{ b_{11}^{(k)} g_1^{(k)}(u) h_1(v) + g_2^{(k)} \sum_{j=1}^{n_{v2}^{(k)}} b_{2j}^{(k)} h_j(v) \right.$$

$$\left. + \sum_{j=1}^{n_{v3}^{(k)}} \sum_{i=3}^{n_u^{(k)}} a_{ij}^{(k)} g_i^{(k)}(u) h_j(v) \right\} \qquad k = 1, 2, \ldots, m \qquad (3.88)$$

However, the derivation of the doublets in Section 3.3.3 is not valid for the general localised junction. It is therefore presented below.

The current-continuity equation for currents flowing out of the general localised junction considered can be written in the form

$$\sum_{k=1}^{m} I_u^{(k)}\{u_1^{(k)}\} = 0 \qquad (3.89)$$

where $I_u^{(k)}\{u_1^{(k)}\}$ is the current flowing out of the junction domain into the kth quadrilateral, i.e.

$$I_u^{(k)}\{u_1^{(k)}\} = \int_{v_1}^{v_2} b_{11}^{(k)} g_1^{(k)}\{u_1^{(k)}\} h_1(v)\, dv \qquad (3.90)$$

Bearing in mind that $g_1^{(k)}\{u_1^{(k)}\} = 1$ and substituting the expression in eqn. 3.90 into eqn. 3.89, the continuity equation is reduced to the equation

$$\sum_{k=1}^{m} b_{11}^{(k)} \int_{v_1}^{v_2} h_1(v)\, dv = 0 \qquad (3.91)$$

By adopting the functions $h_1(v)$ of the same form for all the quadrilaterals, this equation is reduced to eqn. 3.71. The general form of the doublets at the localised junction is then the same as the form of the doublets in eqn. 3.73. However, for the actual interconnection, the Lamé coefficients $e_v^{(1)}$ and $e_v^{(k)}$ have the same form along the interconnection, while for the general localised junction they have not.

3.4.4 *Interconnections of generalised wires*

It is obvious that generalised wires can be interconnected in many ways. There is only one class of junctions where interconnected wires share the common contour and can be analysed in a natural and simple manner. This is the trivial case of the junction obtained by dividing a cylindrical wire into two parts, resulting in two coaxial segments of the same radius. All other junctions of two or more wires, of the same or of different radii and meeting at arbitrary angles, must be analysed by means of the general localised junction model. The current expansions at such interconnections are the same as those described in Section 3.4.3.

3.5 Conclusions

This Chapter has been devoted to the choice of current expansions along generalised wires and over generalised quadrilaterals. To reduce the number of unknowns, current expansions were sought which satisfy the continuity equation at wire ends and junctions, i.e. at generalised quadrilateral edges and interconnections. Polynomial approximation of current was adopted as very convenient for both basic structures, although combined polynomial/trigonometric approximation for generalised wires also seems to be convenient.

It was pointed out that the definitions of entire-domain, almost entire-domain and subdomain expansions are implicitly connected with the electrical size of the structure, and the definitions for the three types of expansion were proposed taking this fact into account. These definitions make almost no distinction between entire-domain and almost-entire-domain expansions, and both were found to be generally superior to subdomain expansions from the computer-storage-requirement viewpoint.

Finally, the interconnections of generalised wires and generalised quadrilaterals which require special attention were considered in more detail and the procedures and logic for treating such interconnections were explained.

Chapter 4

Treatment of excitation

4.1 Introduction

It is possible to excite a metallic structure in many ways. Some examples are shown in Figure 4.1. Sketched in Figure 4.1*a* is the example of a plane wave incident on a metallic structure. The excitation field in this example exists over the whole structure. A similar situation is obtained if a structure is situated in the near field of a source. Figures 4.1*b*–4.1*h* show, schematically, cases of excitation of transmitting structures localised within a small region, and Figure 4.1*i* an example of a more complicated excitation.

All the examples of excitation shown in Figure 4.1 obviously fall into one of the following three categories:

- distributed excitation
- localised excitation
- combined distributed and localised excitation.

Generally speaking, the distributed excitation is obtained if a significant part of the system, or the whole system, is subjected to the excitation field. As mentioned in Section 1.1, the field-analysis method described in this book deals with systems of maximum size in any direction not exceeding several wavelengths. Therefore distributed excitation in our case implies that the excitation field does not vary rapidly in space in any region of the system surface when compared with the system dimensions. Essentially, the excitation of all scatterers we shall consider will be of this type.

The localised excitation is obtained if the extent of the principal part of the impressed electric field, or of impressed currents, is much smaller than the excited system (in any direction) and much smaller than the operating wavelength. (Theoretically, the electromagnetic field of any system of impressed currents extends to infinity, but outside a certain domain it can be neglected. The term 'principal part of the impressed field' refers to the field in that domain.)

Obviously, the localised excitation is typical for transmitting structures. Note, however, that it may also be present in scatterers. For example, if a scatterer has a localised loading (passive or active), it represents a secondary, localised excitation of the structure. This is actually an example of the combined distributed and localised excitation. The combined excitation is also obtained if, in one direction, it can be considered as localised and in the other as distributed. Such an example is shown in Figure 4.1*i*, where a coplanar-strip line is loaded with two sheets representing an antenna. Obviously, the excitation of the antenna is distributed along the line, and localised between the line conductors at any cross-section of the line.

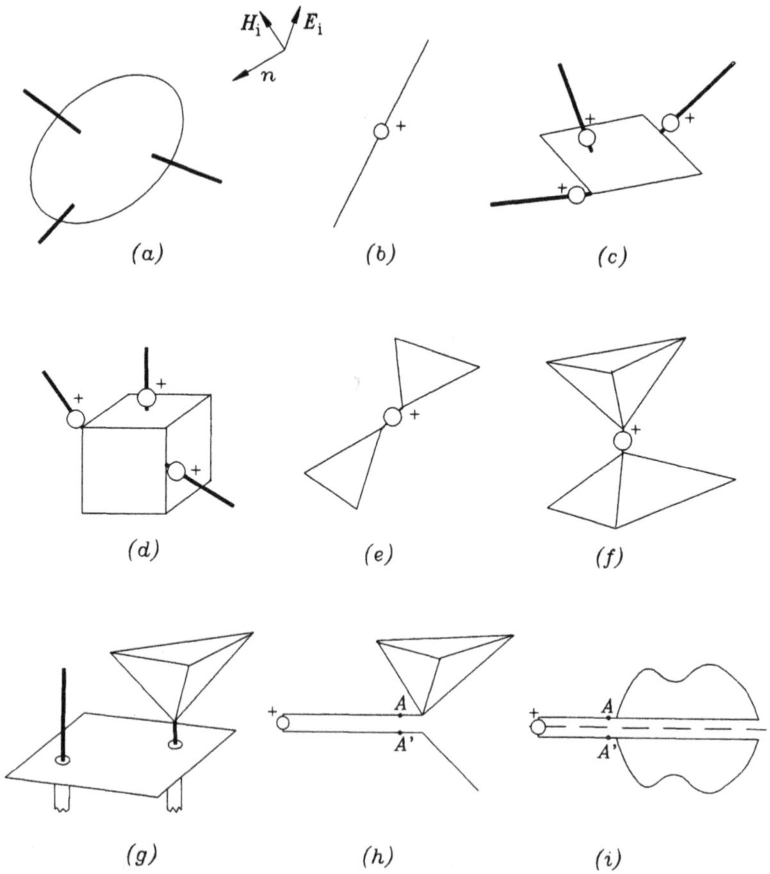

Figure 4.1 *Examples of distributed excitation by incident plane wave (a) and localised excitation (b) to (i)*

This brief Chapter is aimed at deriving the expressions for distributed and localised excitations suitable for the incorporation into equations to be derived in Chapter 5. The combined excitation can in some cases be obtained from these two by superposition.

4.2 Distributed excitation

We shall consider one type of distributed excitation only, that due to an incident plane wave.

Since the source of the plane wave is at a large distance from the structure, there is no coupling between the source of the wave and the structure. We can therefore describe the excitation simply by impressed electric- and magnetic-field vectors $\boldsymbol{E}_i$ and $\boldsymbol{H}_i$. Assume that the wave is incident on the structure from the direction defined by the unit vector $\boldsymbol{n}$, as in Figure 4.2. Let the wave be elliptically polarised, i.e. let the amplitude of the electric-field vector be complex,

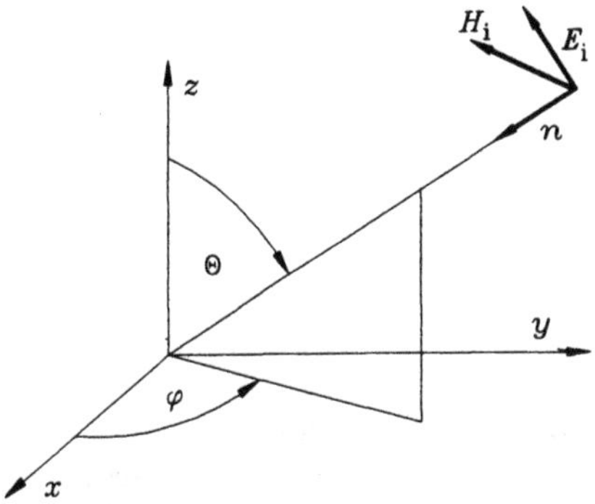

Figure 4.2 Plane wave incident from the direction defined by spherical angles θ and φ

e.g. $\boldsymbol{E}_0$. Finally, let us adopt the co-ordinate origin as the phase reference point. The expression for the (complex) electric field of the wave is then of the form

$$\boldsymbol{E}_i(\boldsymbol{r}) = \boldsymbol{E}_0 \exp(-j\beta \boldsymbol{r} \cdot \boldsymbol{n}) \tag{4.1}$$

where $\beta = \omega\sqrt{(\epsilon\mu)}$ is the phase coefficient,

$$\boldsymbol{r} = x\boldsymbol{i}_x + y\boldsymbol{i}_y + z\boldsymbol{i}_z \tag{4.2}$$

is the position vector in an equiphase plane, and, according to Figure 4.2,

$$\boldsymbol{n} = -\sin\theta\cos\phi\boldsymbol{i}_x - \sin\theta\sin\phi\boldsymbol{i}_y - \cos\theta\boldsymbol{i}_z \tag{4.3}$$

The incident magnetic-field vector of a plane wave is obtained from the incident electric-field vector as

$$\boldsymbol{H}_i(\boldsymbol{r}) = \sqrt{(\epsilon/\mu)}\boldsymbol{n} \times \boldsymbol{E}_i(\boldsymbol{r}) \tag{4.4}$$

4.3 Localised excitations

4.3.1 Introduction

In all transmitting antennas that are of interest in this book, the excitation is of localised type. This implies a convenient model of the generator that can be characterised by the voltage between two electrically close terminals. Although this was not explicitly stated, all the examples of localised excitation in Figures 4.1*b*–4.1*h* were assumed to be of this type.

One of the most important parameters characterising a localised excitation is the impedance as viewed by the generator with respect to its two terminals. If the operating wavelength and the antenna itself are much larger than the distance between the terminals (i.e. at low frequencies), this impedance is not influenced significantly by the actual shape of the terminals. If this is not the case, however, the shape of the terminals may influence the value of the impedance to a considerable degree. Therefore, for example, if the size of the antennas in Figure 4.1*b*–4.1*h* is comparable with the wavelength, one should model excitations individually and as accurately as possible. Such an approach, however, would obviously make the analysis task so diverse that it could become prohibitive.

It is not always simple to adopt two uniquely defined electrically close point-like antenna terminals. For example, in Figures 4.1*h* and 4.1*i* it is not clear where the 2-wire line feeder terminates and the antenna begins. This is because near the feeder physical end the 2-wire line is no longer uniform so that the transmission-line equations, and therefore also the impedance transformation, are not exactly applicable. Only if we assume that the feeder terminates a certain distance from its physical termination, for example at points A and A' in Figures 4.1*h* and 4.1*i*, where higher-order modes due to the line nonuniformity near the end become negligible with respect to the TEM mode, can we make a clear distinction between the feeder (e.g. the part of the line to the left of points A and A' in the two Figures) and the antenna. In the two cases considered, the antenna terminals are the points A and A', and the antenna is the whole structure to the right of these points (thus including also part of the feeder). Of course, we could also consider the whole feeder as part of the antenna, and define the antenna impedance at the generator terminals themselves, but this is obviously usually impractical.

An engineering approach to solving these and related problems is to try to find a few simple models of localised excitation which can be used in all (or at least most) cases and that result in impedance values of acceptable accuracy. Two such generator models have been used widely: the delta-function model (e.g. Reference 11) and the TEM magnetic-current frill model [23, 33]. The first is intended as a rough approximation to any localised excitation. The second is aimed at approximating with relatively high accuracy the coaxial-line excitation for antennas above a perfectly conducting ground plane, as in Figure 4.1*g*. However, we shall see in a later Section that it can also be used as the generator model in cases where there is no ground plane at all. For example, in the analysis of antennas sketched in Figures 4.1*h* and 4.1*i*, both the delta-function generator model and the TEM magnetic-current frill model can be used for the original generator (at the far left of the 2-wire line feeders). Note that, for accurate evaluation of the antenna impedance (as viewed from the antenna terminals), the line must be modelled by appropriate wire segments, for example as illustrated in References 33 and 37.

4.3.2 Delta-function generator

The delta-function generator is a point-like ideal voltage generator. There are several ways in which it can be defined. The simplest is to require that the potential difference between two infinitely close points of generator terminals be equal to the desired EMF of the generator, whatever the shape of the terminals. If the analysis method for wires implies computation of the scalar-potential (e.g. in the case of the so-called 2-potential equation), this type of generator can be represented very simply, requiring that the potential difference between two adjacent points at the interconnection of two wires where the generator is situated (e.g. in Figure 4.1*b*) be equal to that specified. This reasoning can be extended directly to interconnections of wires and plates (Figure 4.1*c*, 4.1*d* and 4.1*g*) or any two bodies meeting at a point (as in Figures 4.1*e* and 4.1*f*).

The delta-function generator poses serious theoretical and, in some instances, numerical problems, if it is assumed to excite cylindrical segments. Consider Figure 4.3, in which a vertical monopole wire antenna above a perfectly

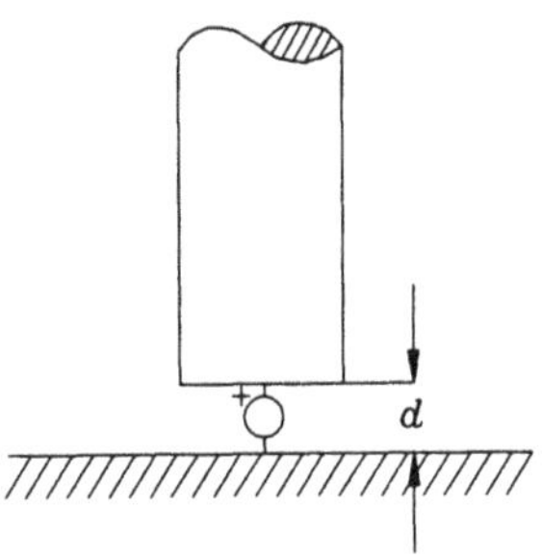

Figure 4.3 Excitation zone of a wire above the ground plane

conducting ground plane is base-driven by such a generator. If the width of the gap d tends to zero (as it should), the capacitance, and therefore also the capacitive current, between the monopole base and the ground become infinitely large. Low-order approximation of current along the monopole usually cannot follow this sharp rise in current intensity in the immediate vicinity of the ground. The imperfect current approximation in this case therefore makes the otherwise meaningless generator model acceptable and, under the circumstances, is even known to yield relatively accurate impedance values. However, for higher-order current approximation the antenna susceptance becomes divergent.

It is interesting that the generalised treatment in this book makes it possible to avoid this difficulty and also to obtain stable values of the antenna impedance for higher-order current approximations. Figure 4.4*a* shows the excitation region represented as two conical segments meeting at a common apex with a delta-function generator in it. The capacitance between the point generator terminals in this case is finite. Thus, by modelling the excitation region properly using the options offered by generalised wires, it is possible to use the delta-function generator without any precautions whatsoever. Figure 4.4*b* shows two flat,

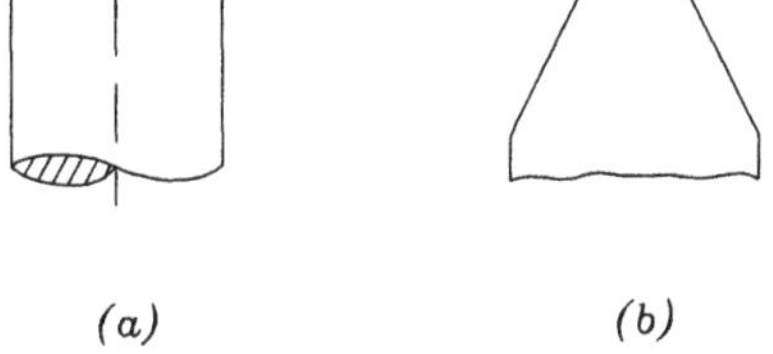

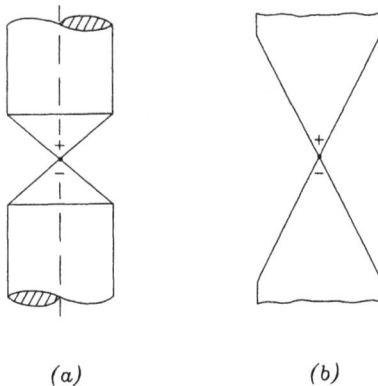

Figure 4.4 Two examples of excitation by delta-function (point) generator which theoretically do not result in infinite antenna susceptance

a Wire antenna with conical terminals
b Flat antenna with triangular terminals

infinitely thin sheets, with analogous modelling of the excitation region which does not lead to infinite antenna susceptance.

Thus, although more sophisticated treatment of the delta-function generator is possible [33], the possibilities of modelling the excitation-region geometry offered by generalised wires and generalised quadrilaterals do not warrant its further discussion. If we encounter any convergence problems in using the delta-function generator and cylinders with flat ends to which the generator is connected, we can always avoid this by modelling the terminals by two short conical segments.

4.3.3 TEM magnetic-current frill

As mentioned, the TEM magnetic-current frill originated as an approximation to coaxial-line excitation. Consider the simplest case possible of a vertical cylindrical monopole antenna of height h and circular cross-section representing a simple protrusion of the inner coaxial-line conductor driving the antenna through an infinite, perfectly conducting ground plane (Figure 4.5*a*). The wave in the line, up to a certain distance below the line opening, is practically a pure TEM wave. Near the opening, however, the line cannot be considered as uniform, and therefore higher-order modes exist in that region of the line in addition to the principal (TEM) mode. Nevertheless, at frequencies for which, approximately, $\beta h < 0.1$, very good results for the antenna admittance are obtained if the electromagnetic field at the coaxial-line opening in Figure 4.5*a* is approximated by the TEM mode only [23, 33]. Assume that this condition is satisfied. The field components in the annular opening of the line are then of the

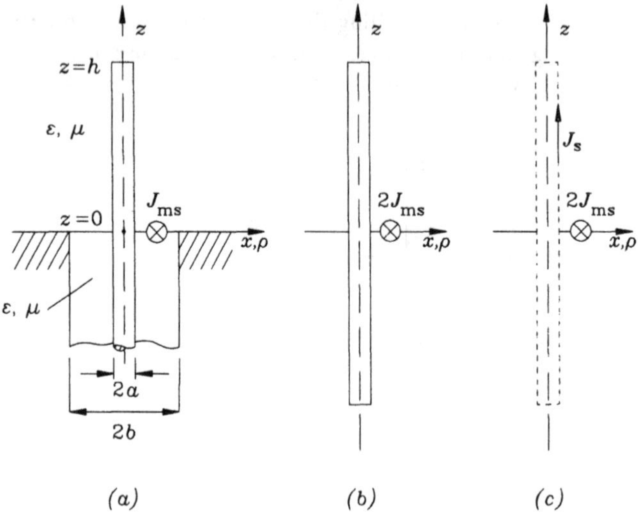

Figure 4.5 Monopole antenna and equivalents

a Monopole antenna above ground plane driven by coaxial line
b Approximate dipole equivalent to the monopole antenna
c Dipole antenna replaced by surface electric currents in a homogeneous medium

form

$$E_\rho(\rho) \simeq E_{\rho\,TEM}(\rho) = \frac{V}{\rho \ln(b/a)} \tag{4.5}$$

$$H_\phi(\rho) \simeq H_{\phi\,TEM}(\rho) = \frac{I(0)}{2\pi\rho} \tag{4.6}$$

where V is the voltage and $I(0)$ the current intensity at the coaxial-line opening, and ρ is the distance of the point in the opening considered from the z axis (see Figure 4.5*a*).

According to the equivalence theorem (e.g. Reference 7), the perfectly conducting ground plane can be extended to cover the coaxial-line opening, provided that an annular layer of surface magnetic currents is placed at the former opening, immediately above the plane, as indicated in Figure 4.5*a*. These magnetic currents are circular, and their density, with respect to the reference direction shown in the Figure, is given by

$$J_{ms\phi}(\rho) = -E_\rho(\rho) \tag{4.7}$$

If the coaxial-line voltage V is assumed to be known, from eqns. 4.5 and 4.7 it follows that $J_{ms\phi}(\rho)$ is also known.

The image theory can be applied to obtain the equivalent system shown in Figure 4.5*b*. In that system the magnetic-current density of the frill is twice that in Figure 4.5*a*. These magnetic currents are the source of the impressed electric field for the symmetrical dipole antenna. Finally, we can apply the equivalence theorem once more to remove the perfectly conducting antenna, which leaves us with the magnetic-current frill and the induced currents and charges on the former antenna surface, situated in a homogeneous medium. A system is thus obtained in which the usual expressions for the retarded potentials can be applied.

The impressed electric field owing to the TEM magnetic-current frill shown in Figure 4.5*c*, at a point $P(x, 0, z)$ lying in the $x0z$ plane, can be shown to have the components [33]

$$E_{ix}(x, z) = -\frac{4zV}{\ln(b/a)} \int_a^b \int_0^\pi \frac{\cos\phi}{r} \frac{dg(r)}{dr} \, d\phi \, d\rho \tag{4.8}$$

$$E_{iz}(x, z) = -\frac{4V}{\ln(b/a)} \int_0^\pi g(r) \Big|_{\rho=a}^{\rho=b} d\phi \tag{4.9}$$

where r is the distance between the source and the field points, ϕ the usual spherical angle co-ordinate, and $g(r)$ is the Green function

$$g(r) = \frac{\exp(-j\beta r)}{r} \tag{4.10}$$

The integrals in eqns. 4.7 and 4.9 can be evaluated numerically with ease, unless the field point P is very close to the magnetic-current frill.

The expressions in eqns. 4.8 and 4.9 are important when analysing general structures excited by a coaxial line. If a vertical monopole only is considered, as in Figure 4.5*a*, we shall see that we need solely the E_{iz} component along the monopole axis (the z axis in Figure 4.5), because along the axis the other component is zero. This expression can be obtained from eqn. 4.9, noting that for points along the monopole axis $r = (\rho^2 + z^2)^{1/2}$. The result is

$$E_{iz}(z) = \frac{4\pi V}{\ln(b/a)} \{g(r_a) - g(r_b)\} \tag{4.11}$$

where

$$r_a = (a^2 + z^2)^{1/2} \qquad r_b = (b^2 + z^2)^{1/2} \tag{4.12}$$

It can be shown in a similar manner that the impressed magnetic field due to the TEM magnetic-current frill shown in Figure 4.5*c*, at a point $P(x, 0, z)$ lying in the $x0z$ plane, has the y component only, that is computed from the formula

$$H_{iy}(x, z) = \frac{j4\omega\epsilon_0 V}{\ln(b/a)} \int_a^b \int_0^\pi \cos\phi \, g(r) \, d\phi \, d\rho \tag{4.13}$$

It should be mentioned at this point that, as a consequence of the annular magnetic-current-frill excitation, the current derivative has a discontinuity at the frill location. The value $(dI/dz)_{z=0+}$ of the derivative is uniquely determined by the coaxial-line dimensions and the voltage V at the line opening, as demonstrated below. Although it has been mentioned that in the approach adopted in this book it is not necessary to take this condition into account, in some numerical solutions for current distribution when using the TEM magnetic-current frill generator it can assist in obtaining more stable and accurate results.

We shall consider the case in Figure 4.5*a*, with the opening closed by the frill and the line filled with perfect conductor. To determine the value of $(dI/dz)_{z=0+}$, note that at the antenna surface in Figure 4.5*a* the electric-field vector at $z = 0+$ has only the radial component E_ρ that is tangential to the magnetic-current frill. According to boundary conditions and referring to Figure 4.5*a*,

$$\begin{aligned} E_\rho(\rho = a, z = 0+) - E_\rho(\rho = a, z = 0-) \\ = E_\rho(\rho = a, z = 0+) = -J_{ms\phi}(\rho = a) \end{aligned} \tag{4.14}$$

because $E_\rho(\rho = a, z = 0-) = 0$ (the point is inside the perfect conductor). The antenna charge per unit length is related to the electric field as

$$Q'(z) = 2\pi\epsilon a E_\rho(\rho = a, z) \tag{4.15}$$

The first current derivative and this charge are interrelated through the continuity equation, i.e.

$$\frac{dI(z)}{dz} = -j\omega Q'(z) \tag{4.16}$$

From eqns. 4.5, 4.7 and 4.14–4.16, it follows that

$$\left.\frac{\mathrm{d}I(z)}{\mathrm{d}z}\right|_{z=0+} = -j\,\frac{2\pi\epsilon\omega V}{\ln(b/a)} = -j\beta Y_c V \tag{4.17}$$

where Y_c is the characteristic admittance of the coaxial line (if the frill approximates actual coaxial-line excitation). Note that the expression on the right-hand side of eqn. 4.17 is imaginary. This means that the active current component (that in phase with the voltage) has zero derivative with respect to z at $z = 0+$, and the reactive current component must have the derivative at $z = 0+$ as in this equation.

The condition in eqn. 4.17 can easily be incorporated into any smooth approximation of the antenna current distribution, such as the polynomial approximation, resulting in very stable and accurate results for current and charge distribution in the vicinity of the magnetic-current frill. However, these distributions are very rapidly varying along the z axis in the excitation region, and special precautions must be taken to provide a sufficiently accurate approximation for these distributions near the magnetic-current frill. Note that, if the Galerkin method is used, accurate results for current distribution are also obtained without taking this condition into account.

The TEM magnetic-current frill is a very convenient excitation model for monopole antennas driven by coaxial lines above large ground planes. Obviously, however, it is not clear how it could be used in cases where a large plane is not present, for example those sketched in Figure 4.1*c* and 4.1*d*. In addition, it is of interest to analyse the possibility of using the TEM magnetic-current-frill excitation in other cases, for example those sketched in Figures 4.1*f*–4.1*i*.

In particular, if the frill is placed over a finite metal plate, so that image theory cannot be used, the problem becomes quite intricate, for the following reasons. The frill (surface magnetic currents) produces a discontinuous distribution of the excitation field over the plate: the intensity of the electric-field-vector component tangential to the plate equals half the intensity of the surface magnetic currents for points below the frill, and is zero elsewhere. Consequently, the electric-charge distribution and the first derivative of the electric surface current are also discontinuous over the plate. Therefore the analysis of the structure, e.g. using the method of moments, tends to be quite complicated.

To be more specific, modelling of a simple wire-to-plate junction (i.e. without excitation) is done usually as proposed in either Reference 46 or 84. In the first case, modelling is performed by using a wire segment, a surface patch and the so-called attachment mode, i.e. a superimposed patch with a special current distribution. In the second case, modelling is performed by using a wire segment and a minimum of three surface patches, but with the same current distribution as for any other patch. If a magnetic current frill exists at the junction, however, this becomes still more complicated, because additional attachment modes or surface patches are needed for modelling the junction. Therefore it would be very convenient if an excitation along the wire could be found that was equivalent to the frill excitation.

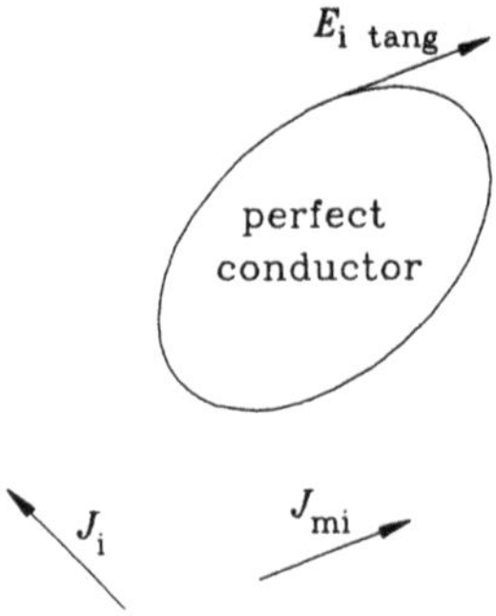

Figure 4.6 Perfectly conducting body in the field of impressed electric and magnetic currents

4.3.4 A theorem on transfer of excitation

To solve the problems mentioned in the last paragraph, we need a simple theorem which justifies the transfer of excitation of any antenna from the original excitation region to the antenna surface. Consider the system shown in Figure 4.6, where impressed electric and magnetic currents are exciting an arbitrary perfectly conducting body. By definition, impressed currents are analogous to ideal current generators of electrical networks. Therefore, the secondary field owing to induced charges and currents does not influence the impressed currents. As a consequence, the total field is that of the impressed currents, plus the field of the induced currents and charges on the surface of the body.

If we deal with transmitting antennas, the field of impressed currents (that are approximately equivalent to the actual excitation) in the radiation zone is negligible with respect to the field of the induced charges and currents. We are therefore interested in the impressed currents only so far as the excitation of the antenna itself is concerned. The antenna being perfectly conducting, what we need to know is the tangential component of the electric-field vector $\boldsymbol{E}_{i\ tang}$, or the tangential component of the magnetic-field vector $H_{i\ tang}$, on the antenna surface due to the impressed currents, and not the currents themselves. Thus, the equivalent excitation to the impressed currents is the tangential electric or magnetic field (i.e. the impressed field) on the antenna surface due to these currents, and the actual excitation can be removed.

As a result of this simple theorem, we can, in fact, assume any convenient distribution of the impressed electric field on the antenna surface as the excitation of the antenna. For a TEM magnetic-current frill, e.g. that shown in Figure 4.5*b*, we can remove the frill and consider the axial electric field of the frill along the wire antenna as the impressed electric field. If this field is concentrated sufficiently that the voltage between two close terminals of the antenna can be defined, the transformed excitation has all the properties it should have.

Note that the impressed field must satisfy Maxwell's equations, since if it does not, it will not be possible for charges and currents over the antenna surface to compensate this field and make the total field zero everywhere inside the antenna body. This condition, however, is certainly satisfied if the impressed field is identical to the field of a system of impressed currents, e.g. of a TEM magnetic-current frill. Therefore, we can use the TEM magnetic-current frill in a much broader sense than just for the analysis of antennas above an infinite,

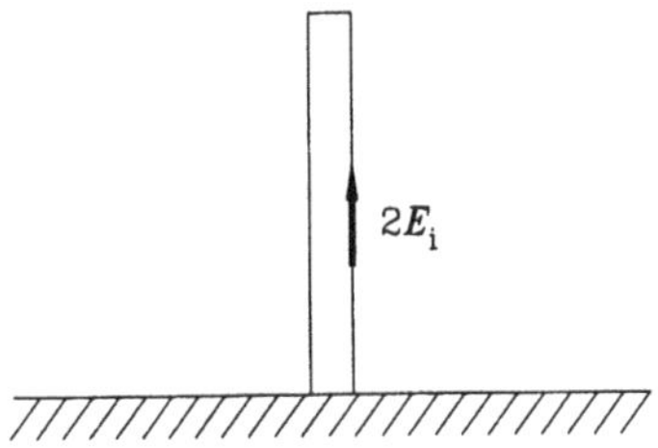

Figure 4.7 Monopole antenna from Figure 4.5a with excitation transferred from the frill to the monopole

E_i is the electric field due to the frill itself (considered in a vacuum)

perfectly conducting ground plane and driven by a coaxial line. Of course, no physical interpretation of the impressed field in such cases will be possible, but, under these circumstances, it will permit the analysis of the transmitting antenna driven by a generator. We may define the voltage of the equivalent generator in the same way as above, adopt any convenient ratio b/a and then analyse the transmitting antenna using, in fact, a convenient TEM magnetic-current-frill model of the excitation transferred (in the above sense) to the antenna surface.

The most important case in practice is probably that of a monopole antenna connected to a metal plate of finite size, possibly near its edge. Consider first the monopole antenna above a perfectly conducting ground plane, shown in Figure 4.5*a*, excited by a magnetic current frill. The above theorem states that the frill can be removed and the excitation field transferred along the monopole and to the ground plane (where it exists). From Figure 4.5*b*, it is obviously equal to twice the electric field of the frill proper (if the frill were situated in a vacuum, i.e. with no ground plane present). This means that, for the monopole above a ground plane, the excitation can be transferred only along the monopole, with double intensity, as indicated in Figure 4.7. (Of course, the total field is not the same, since the frill is removed; as explained, however, this is usually of no importance.)

Consider now a finite metal plate instead of an infinite ground plane. The above solution is no longer exact, but we shall see that it nevertheless also yields remarkably accurate results for plates of finite size, even if the wire is close to the plate edge.

In the examples presented in later Chapters, both types of universal generator model will be used: the delta-function generator, and the magnetic-current frill or the impressed electric field owing initially to a TEM magnetic-current frill, but transferred to the driven wire for convenience.

4.3.5 *Examples of application of the theorem on transfer of excitation*

The simplest, delta-function generator can often be used in the analysis of wire and plate structures, in fact in all cases when the wire radius is much less than its height and much less than the wavelength at the operating frequency. (If the wire radius is much less than the wire height, as mentioned, a low-order approximation of current along the wire cannot usually reflect the infinite capacitive current implied by the delta-function generator. If the wire radius is

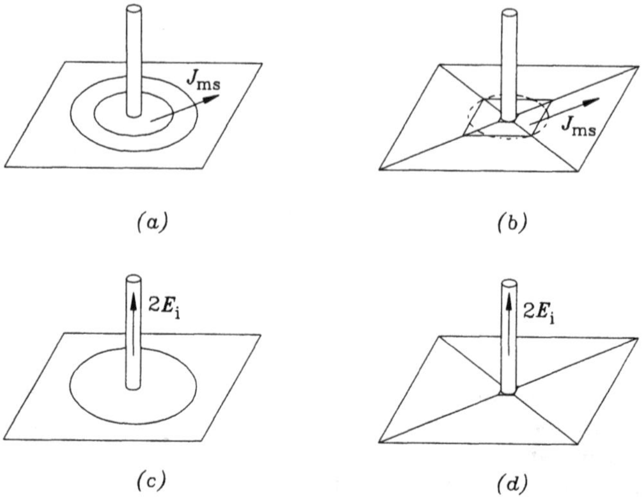

Figure 4.8 Sketches of various approximations of geometry of a junction of plate and wire

The junction is modelled by one wire and (*a*) one surface patch and two attachment modes, (*b*) eight surface patches, (*c*) one surface patch and one attachment mode, and (*d*) four surface patches

much smaller than the wavelength at the operating frequency, the wire-antenna impedance is not influenced significantly by the actual shape of the generator terminals.)

If the wire is electrically relatively thick, however, more accurate modelling of the excitation zone is needed. If the electrical radius of the outer coaxial-feeder conductor is less than about one-tenth of the wavelength, the approximation of the coaxial-line excitation by the TEM magnetic-current frill appears to be sufficiently accurate [33]. However, as mentioned, direct solution of the structure with the magnetic-current frill is rather complicated.

For example, consider a wire connected to the central part of a rectangular plate and excited by the magnetic current frill. The junction can be modelled by one wire segment, one surface patch and two attachment modes (one for the frill and one for rapidly varying current near the junction), as shown in Figure 4.8*a*. Alternatively, it can be modelled by one wire segment and eight surface patches (four for the frill and four for the rest of the plate), as shown in Figure 4.8*b*.

However, the principal impressed field of the frill is localised within a domain of a few outer radii of the coaxial line that the frill is approximating. Therefore, if the frill and the wire are not close to the plate edge, the approximate form of the above theorem applies, the frill excitation can be transferred to the wire, and the frill removed. The additional attachment mode or additional partitioning required by the frill therefore becomes unnecessary, and the analysis becomes essentially the same as if the excitation were a delta-function generator or a plane wave, i.e. the simplest possible. The results for the current distribution over the body, however, are the same as in the actual case.

Consider again the wire connected to the central part of a rectangular plate excited by the magnetic-current frill. Once the transfer of excitation has been performed, the junction can be modelled by one wire segment, one surface patch and one attachment mode, as sketched in Figure 4.8*c*. Alternatively, it can be modelled by one wire segment and four surface patches, as indicated in Figure 4.8*d*. The latter approach is obviously simpler. In addition, which may be even more important, it can be applied equally to excitations at the centre of the plate and near to or at plate discontinuities, while the first approach requires separate attachment modes to be devised in all these cases [58, 59].

4.4 Conclusions

This Chapter was aimed at introducing convenient models of excitation for antennas. The distributed excitation, that of a uniform plane wave, was introduced first. The localised excitation was considered next. It was concluded that practical reasons dictate the use of a few simple models of localised excitation, which can be used in most cases; the delta-function generator is one such possibility, and the TEM magnetic-current frill (usually changed to an approximately equivalent distribution of impressed electric field) is the other.

An important addition to the standard treatment of excitation relates to relatively thick monopole antennas connected to metal plates of finite size and driven at the junction by a magnetic-current frill approximating the coaxial-line excitation. By a simple theorem the frill excitation is transferred to the wire, which enables the plate and the monopole to be analysed in the same manner as if the excitation were by a delta-function generator or a plane wave, i.e. an additional attachment mode or finer partitioning of the plate is avoided.

Chapter 5

Electromagnetic field of currents over generalised surface elements

5.1 Introduction

This Chapter is devoted to derivation of expressions for the electromagnetic fields due to known distributions of currents over arbitrary curvilinear rectangles, over bilinear surfaces, and along generalised wires and truncated cones. The first step is determination of the Lorentz potentials, from which the electric- and magnetic-field vectors in the near and far zones are then evaluated. With these expressions for the near-zone field vectors, the method for determining the current distribution will be presented in Chapter 6 based on the Galerkin procedure.

5.2 Electromagnetic field of currents over curvilinear quadrilaterals and bilinear surfaces

5.2.1 *General expression for potentials and field vectors*

Consider a system of time-harmonic surface charges of density ρ_s and surface currents of density $\boldsymbol{J}_s$ distributed over surfaces of perfectly conducting bodies situated in a vacuum. Inside perfectly conducting bodies there is no field. Therefore we can assume that currents and charges exist in a vacuum over geometrical surfaces, subject to their distribution being such that boundary conditions on these surfaces are the same as for perfectly conducting bodies. Therefore the medium can be considered as homogeneous, and the Lorentz scalar potential V and vector potential $\boldsymbol{A}$ can be calculated in the known way:

$$V = \frac{1}{\epsilon_0}\int_S \rho_s g(R)\,\mathrm{d}S \tag{5.1}$$

$$\boldsymbol{A} = \mu_0 \int_S \boldsymbol{J}_s g(R)\,\mathrm{d}S \tag{5.2}$$

In these expressions, $g(R)$ is the free-space Green function

$$g(R) = \frac{\exp(-j\beta R)}{4\pi R} \tag{5.3}$$

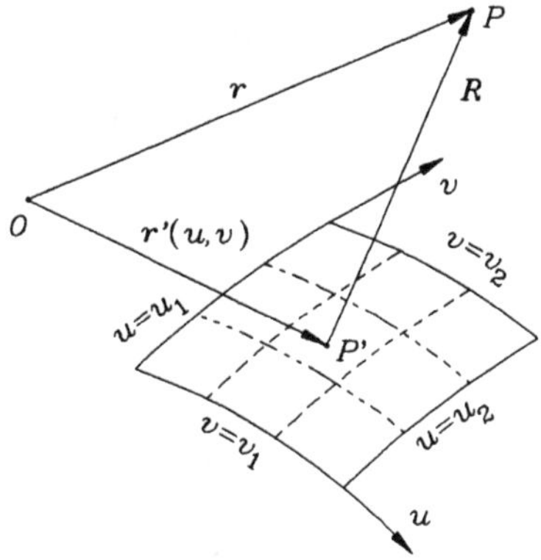

Figure 5.1 Curvilinear quadrilateral with local (u, v) co-ordinates, source point (P') and field point (P)

β is the free-space propagation coefficient

$$\beta = \omega\sqrt{(\epsilon_0\mu_0)} \tag{5.4}$$

and R is the distance of the field point P from the source point P'

$$\boldsymbol{R} = \boldsymbol{r}' - \boldsymbol{r} \qquad R = |\boldsymbol{R}| \tag{5.5}$$

where $\boldsymbol{r}$ is the position vector of the field point, and $\boldsymbol{r}' = \boldsymbol{r}'(u, v)$ is the position vector of the source point, as shown in Figure 5.1 for a surface in the form of a generalised quadrilateral.

The electric- and magnetic-field vectors are determined from the potentials by means of the relations

$$\boldsymbol{E} = -j\omega\boldsymbol{A} - \text{grad } V \tag{5.6}$$

$$\boldsymbol{H} = \frac{1}{\mu_0} \text{curl } \boldsymbol{A} \tag{5.7}$$

Substituting the expressions in eqns. 5.1 and 5.2 for the potentials into eqns. 5.6 and 5.7 and making use of the continuity equation

$$\rho_s = (j/\omega)\,\text{div}_s\boldsymbol{J}_s \tag{5.8}$$

the electric- and magnetic-field vectors can be expressed in terms of surface currents alone. Note that the gradient and curl operators operate on the field-point co-ordinates. They can therefore be introduced under the integrals and made to operate on the Green function alone. This yields

$$\text{grad}\{\text{div}_s\boldsymbol{J}_s g(R)\} = \text{div}_s\boldsymbol{J}_s\,\text{grad}\,g(R) = \text{div}_s\boldsymbol{J}_s\,\frac{\text{d}g(R)}{\text{d}R}\,\boldsymbol{i}_R \tag{5.9}$$

and

$$\text{curl}\{\boldsymbol{J}_s g(R)\} = \text{grad}\{g(R)\} \times \boldsymbol{J}_s = \frac{\text{d}g(R)}{\text{d}R}\,\boldsymbol{i}_R \times \boldsymbol{J}_s \tag{5.10}$$

where $\boldsymbol{i}_R$ is the unit vector directed from the source point to the field point, $\boldsymbol{i}_R = \boldsymbol{R}/R$.

The final expressions thus obtained for the electric- and magnetic-field vectors are

$$\boldsymbol{E} = -j\omega\mu_0 \left\{ \int_S \boldsymbol{J}_s g(R)\,\mathrm{d}S + \frac{1}{\beta^2} \int_S \mathrm{div}_s \boldsymbol{J}_s\, \boldsymbol{i}_R \frac{\mathrm{d}g(R)}{\mathrm{d}R}\,\mathrm{d}S \right\} \qquad (5.11)$$

and

$$\boldsymbol{H} = \int_S \frac{\mathrm{d}g(R)}{\mathrm{d}R}\, \boldsymbol{i}_R \times \boldsymbol{J}_s\,\mathrm{d}S \qquad (5.12)$$

Note that in addition to the Green function $g(R)$, in the integrals we have only the derivative of the Green function with respect to R.

There is an important fact relating to the last two expressions when used for formulating boundary conditions on the (perfect) conductor surface. This fact is essential for proper formulation of the integral equations for current distribution, as will be explained in what follows.

Boundary conditions are usually formulated for points that are either just outside S or just inside S. Theoretically, the expressions in eqns. 5.11 and 5.12 yield the correct field values in the two cases. However, when the expressions in eqns. 5.11 and 5.12 are evaluated numerically, we cannot take into account these fine details. Rather, we are able to compute the field vectors only by assuming the field point to be on S itself. This means, however, that we do not obtain the correct values of the field vectors we need (just outside S or just inside S). The correct field values are obtained only if we introduce an additional term for the normal component of the electric-field vector and an additional term for the tangential component of the magnetic-field vector. The tangential component of the electric-field vector and the normal component of the magnetic-field vector require no correction. This can be understood by a closer inspection of the expressions in eqns. 5.11 and 5.12 as follows.

Consider first the electric-field vector in eqn. 5.11. Assume that the field point is on the geometrical surface S of the conducting body, where we imagine the surface currents and charges. We can partition the surface S into a small circle around the field point, and the rest of the surface, and integrate over these two surfaces separately.

Consider those parts of the integrals in eqn. 5.11 which are evaluated over the surface of the small circle. Since local $\boldsymbol{J}_s$ and $\boldsymbol{i}_R$ are then tangential to the surface, so is the electric-field vector. Therefore, according to eqn. 5.11, surface charges and currents on the small circle produce only the tangential component of the electric-field vector at the field point. This is not a result we can use to formulate the boundary condition for the normal component of the electric-field vector, for we know from quasistatic analysis that exactly one-half of the normal component of the electric-field vector at the field point that is just outside S is due to the surface charge on the small circle, the other half being due to all the other sources. Just inside S, these two normal components are of equal magnitude and in opposite directions.

Therefore, at points infinitely close to conducting surfaces with surface charges and currents, the correct normal component of the electric-field vector is obtained if to the expression in eqn. 5.11, calculated as if the field point were on the very surface, we add $\pm(\rho_s/2\epsilon)\boldsymbol{n}$. Here ρ_s is the local surface-charge density and $\boldsymbol{n}$ is the outward unit vector normal to S at the field point. (The plus sign is used if the field point is imagined just outside S, and the minus sign if it is imagined just inside S.) The boundary condition requires that this sum, plus the normal component of the impressed electric-field vector, be equal to $(\rho_s/\epsilon)\boldsymbol{n}$ if the field point is just outside S, and zero if it is just inside S. The same equation therefore results in both cases (if the field point is just outside and just inside S). It is obtained by equating the sum of the right-hand side in eqn. 5.11 and the impressed electric-field vector to $+(\rho_s/2\epsilon)\boldsymbol{n}$.

Similar reasoning applies for the expression in eqn. 5.12. Let us partition the surface S again into a small circle around the field point and the rest of S, and concentrate on the field due to currents on the small circle. It is a simple matter to understand that these currents can produce only the component of $\boldsymbol{H}$ that is locally normal to S. On the other hand, from quasistatic relations we know that one-half of the total tangential component of the magnetic-field vector at the point considered, just outside S, equal to $1/2\boldsymbol{J}_s \times \boldsymbol{n}$, is due to the local surface current, and the other half to the rest. Just inside S, these two components are of equal magnitude and in opposite directions. Therefore, the correct expression for the magnetic-field vector at points infinitely close to S is obtained if $\pm 1/2\boldsymbol{J}_s \times \boldsymbol{n}$ is added on the right-hand side of eqn. 5.12. (The plus and minus signs have the same meaning as for the normal component of the electric-field vector.) The boundary condition for the tangential component of the magnetic-field vector requires that this sum, plus the tangential component of the impressed magnetic-field vector, be equal to $+\boldsymbol{J}_s \times \boldsymbol{n}$ if the field point is just outside S, and zero if it is just inside S. The same equation results therefore in the two cases (for the field point just outside and just inside S). It is obtained by equating the sum of the right-hand side in eqn. 5.12 and the impressed magnetic-field vector to $+1/2\boldsymbol{J}_s \times \boldsymbol{n}$.

For far-field points the last two expressions can be simplified. If $|\boldsymbol{R}| \gg |\boldsymbol{r}'|$ (point in the far zone), we have

$$\boldsymbol{R} \simeq \boldsymbol{r} - \boldsymbol{r}' . \boldsymbol{i}_R \qquad R = |\boldsymbol{R}| \simeq |\boldsymbol{r}| = r \tag{5.13}$$

so that

$$g(R) = \frac{\exp(-j\beta R)}{R} \simeq \left\{\frac{\exp(-j\beta r)}{r}\right\} \exp(j\beta \boldsymbol{r}' . \boldsymbol{i}_R) \tag{5.14}$$

If, in addition, $\beta \gg 1/R$, then

$$\begin{aligned}\frac{\mathrm{d}g(R)}{\mathrm{d}R} &= \frac{\mathrm{d}}{\mathrm{d}R}\left\{\frac{\exp(-j\beta R)}{R}\right\} = -\left(j\beta + \frac{1}{R}\right)\frac{\exp(-j\beta R)}{R} \\ &\simeq -j\beta\,\frac{\exp(-j\beta R)}{R} \simeq \left\{-j\beta\,\frac{\exp(-j\beta r)}{r}\right\} \exp(j\beta \boldsymbol{r}' . \boldsymbol{i}_R)\end{aligned} \tag{5.15}$$

To obtain the far-zone electric-field vector, it is not necessary, however, to substitute all these expressions into eqn. 5.11. It is well known that in the far zone the electric-field vector of any antenna or scatterer can be expressed as

$$\boldsymbol{E} = -j\omega \boldsymbol{A}_{\text{normal to } R} \tag{5.16}$$

where $\boldsymbol{A}_{\text{normal to } R}$ is the component of the magnetic vector potential normal to the direction of unit vector $\boldsymbol{i}_R$,

$$\boldsymbol{A}_{\text{normal to } R} = \boldsymbol{i}_R \times (\boldsymbol{A} \times \boldsymbol{i}_R) \tag{5.17}$$

Using eqns. 5.2, 5.14 and 5.16 this results in

$$\boldsymbol{E} = -j\omega\mu_0 \frac{\exp(-j\beta r)}{r} \boldsymbol{i}_R \times \left\{ \int_S \boldsymbol{J}_s \exp(j\beta \boldsymbol{r}' . \boldsymbol{i}_R) \, \mathrm{d}S \times \boldsymbol{i}_R \right\} \tag{5.18}$$

since the term in curly brackets on the right-hand side of eqn. 5.14 does not depend on the source point co-ordinates, over which the integration is performed, and can therefore be extracted outside the integral.

The magnetic-field vector is obtained from eqns. 5.12 and 5.15:

$$\boldsymbol{H} = -j\beta \frac{\exp(-j\beta r)}{r} \boldsymbol{i}_R \times \int_S \boldsymbol{J}_s \exp(j\beta \boldsymbol{r}' . \boldsymbol{i}_R) \, \mathrm{d}S \tag{5.19}$$

Note that in the far zone

$$\boldsymbol{H} = (\epsilon_0/\mu_0)^{1/2} \boldsymbol{i}_R \times \boldsymbol{E} \tag{5.20}$$

as it should be.

Starting from these expressions, it is possible to calculate the electric- and magnetic-field vectors due to any known surface currents existing over an arbitrary surface. We are interested, however, in computing the field vectors for currents in the form of arbitrary basis functions existing over generalised quadrilaterals and along generalised wires.

5.2.2 Potentials and field vectors of currents over generalised quadrilaterals

Assume that all system surfaces are approximated by generalised quadrilaterals. Assume, in addition, that for all the quadrilaterals the local (u, v) co-ordinates are defined, as in Figure 5.1. As explained in Section 3.3, we represent the surface-current-density vector with respect to these co-ordinates. The area element $\mathrm{d}S$ of the quadrilateral is given in eqn. 2.10. It was explained that we need to consider only the u component J_{su} of the surface-current-density vector, because the total current can be represented as two (or three) overlapping u components. It has also been explained that it is convenient to represent J_{su} in terms of an auxiliary function $\partial I_u/\partial v$, as in eqn. 3.59. With this in mind, and using the expression in eqn. 3.56 for the divergence of $\boldsymbol{J}_s$, after simple manipulations we first obtain the following expressions for the scalar and vector potentials:

$$V = \frac{j}{\omega\epsilon_0} \int_{v_1}^{v_2} \int_{u_1}^{u_2} \frac{\partial}{\partial u} \left(\frac{\partial I_u}{\partial v} \right) g(R)\, \mathrm{d}u\, \mathrm{d}v \tag{5.21}$$

$$\boldsymbol{A} = \mu_0 \int_{v_1}^{v_2} \int_{u_1}^{u_2} \left(\frac{\partial I_u}{\partial v} \right) \frac{\partial \boldsymbol{r}'}{\partial u} g(R)\, \mathrm{d}u\, \mathrm{d}v \tag{5.22}$$

The electric-field vector in this representation is given by

$$\boldsymbol{E} = -j\omega\mu_0 \left\{ \int_{v_1}^{v_2} \int_{u_1}^{u_2} \left(\frac{\partial I_u}{\partial v} \right) \frac{\partial \boldsymbol{r}'}{\partial u} g(R)\, \mathrm{d}u\, \mathrm{d}v + \frac{1}{\beta^2} \int_{v_1}^{v_2} \int_{u_1}^{u_2} \frac{\partial}{\partial u} \left(\frac{\partial I_u}{\partial v} \right) \boldsymbol{i}_R \frac{\mathrm{d}g(R)}{\mathrm{d}R}\, \mathrm{d}u\, \mathrm{d}v \right\} \tag{5.23}$$

and the magnetic-field vector is calculated from the expression

$$\boldsymbol{H} = \int_{v_1}^{v_2} \int_{u_1}^{u_2} \frac{\mathrm{d}g(R)}{\mathrm{d}R} \boldsymbol{i}_R \times \frac{\partial \boldsymbol{r}'}{\partial u} \left(\frac{\partial I_u}{\partial v} \right) \mathrm{d}u\, \mathrm{d}v \tag{5.24}$$

It was explained in Section 3.3.3 that it was convenient to represent not J_{su}, but rather $\partial I_u/\partial v$, in terms of the basis functions $f_i(u)$ and $h_j(v)$, as in eqn. 3.60. With this representation for $\partial I_u/\partial v$, the scalar potential due to the u component of surface-current-density vector can be expressed as

$$V = \sum_{j=1}^{n_v} \sum_{i=1}^{n_u} a_{ij} V_{ij} \tag{5.25}$$

where the partial scalar potentials, due to pairs $f_i(u)$ and $h_j(v)$ of the basis functions, have the form

$$V_{ij} = \frac{j}{\omega\epsilon_0} \int_{v_1}^{v_2} \int_{u_1}^{u_2} \frac{\mathrm{d}f_i(u)}{\mathrm{d}u} h_j(v) g(R)\, \mathrm{d}u\, \mathrm{d}v \tag{5.26}$$

and a_{ij} are unknown coefficients in the expansion for $\partial I_u/\partial v$ that need to be determined.

In a similar manner, the vector potential can be written as

$$\boldsymbol{A} = \sum_{j=1}^{n_v} \sum_{i=1}^{n_u} a_{ij} \boldsymbol{A}_{ij} \tag{5.27}$$

where the partial vector potentials, due to pairs $f_i(u)$ and $h_j(v)$ of the basis functions, are given by

$$A_{ij} = \mu_0 \int_{v_1}^{v_2} \int_{u_1}^{u_2} f_i(u) h_j(v) \frac{\partial \boldsymbol{r}'}{\partial u} g(R)\, du\, dv \tag{5.28}$$

With these expressions for the potentials, the electric-field vector due to the u component of the surface-current-density vector and the adopted current expansion over generalised quadrilaterals takes the form

$$\boldsymbol{E} = \sum_{j=1}^{n_v} \sum_{i=1}^{n_u} a_{ij} \boldsymbol{E}_{ij} \tag{5.29}$$

where

$$\boldsymbol{E}_{ij} = -j\omega \boldsymbol{A}_{ij} - \frac{j\omega\mu_0}{\beta^2} \int_{v_1}^{v_2} \int_{u_1}^{u_2} \frac{df_i(u)}{du} h_j(v) \boldsymbol{i}_R \frac{dg(R)}{dR}\, du\, dv \tag{5.30}$$

Similarly, the magnetic-field vector can be written as

$$\boldsymbol{H} = \sum_{j=1}^{n_v} \sum_{i=1}^{n_u} a_{ij} \boldsymbol{H}_{ij} \tag{5.31}$$

where

$$\boldsymbol{H}_{ij} = \int_{v_1}^{v_2} \int_{u_1}^{u_2} \frac{dg(R)}{dR} \boldsymbol{i}_R \times \frac{\partial \boldsymbol{r}'}{\partial u} f_i(u) h_j(v)\, du\, dv \tag{5.32}$$

For far-field points, the electric-field vector is obtained from eqns. 5.16, 5.17 and 5.27, with partial vector potentials given by

$$\boldsymbol{A}_{ij} = \mu_0 \frac{\exp(-j\beta r)}{r} \int_{v_1}^{v_2} \int_{u_1}^{u_2} f_i(u) h_j(v) \frac{\partial \boldsymbol{r}'}{\partial u} \exp(j\beta \boldsymbol{r}' . \boldsymbol{i}_R)\, du\, dv \tag{5.33}$$

and the magnetic-field vector is then obtained from eqn. 5.20.

5.2.3 *Potentials and field vectors of polynomial distribution of currents over bilinear surfaces*

Consider now the special, but important, case when $f_i(u) = u^{i-1}$, $h_j(v) = v^{j-1}$ and the current exists over a bilinear surface, described in eqn. 2.28. The expressions in eqns. 5.26, 5.28, 5.30 and 5.32 for the partial potentials and partial field vectors in that case become

$$V_{ij} = \frac{j}{\omega\epsilon_0} (i-1) P_{i-1,\, j} \tag{5.34}$$

$$\boldsymbol{A}_{ij} = \mu_0 (\boldsymbol{r}_u P_{ij} + \boldsymbol{r}_{uv} P_{i,\, j+1}) \tag{5.35}$$

$$\boldsymbol{E}_{ij} = -j\omega\mu_0 \Bigg[\boldsymbol{r}_u P_{ij} + \boldsymbol{r}_{uv} P_{i,j+1} + \frac{1}{\beta^2}(i-1)\Big\{ (\boldsymbol{r}-\boldsymbol{r}_c)Q_{i-1,j} - \boldsymbol{r}_u Q_{ij} - \boldsymbol{r}_v Q_{i-1,j+1} - \boldsymbol{r}_{uv} Q_{i,j+1} \Big\} \Bigg] \quad (5.36)$$

$$\boldsymbol{H}_{ij} = (\boldsymbol{r}-\boldsymbol{r}_c)\times \boldsymbol{r}_u Q_{ij} + \{\boldsymbol{r}_u \times \boldsymbol{r}_v + (\boldsymbol{r}-\boldsymbol{r}_c)\times \boldsymbol{r}_{uv}\} Q_{i,j+1} - \boldsymbol{r}_v \times \boldsymbol{r}_{uv} Q_{i,j+2} \quad (5.37)$$

where

$$P_{ij} = \int_{v_1}^{v_2}\int_{u_1}^{u_2} u^{i-1} v^{j-1} g(R)\, \mathrm{d}u\, \mathrm{d}v \quad (5.38)$$

and

$$Q_{ij} = \int_{v_1}^{v_2}\int_{u_1}^{u_2} u^{i-1} v^{j-1} \frac{1}{R}\frac{\mathrm{d}g(R)}{\mathrm{d}R}\, \mathrm{d}u\, \mathrm{d}v \quad (5.39)$$

The far-field approximations in this case are obtained if $g(R)$ in eqn. 5.38 is replaced by the approximate expression in eqn. 5.14, which results in

$$P_{ij} = \frac{\exp(-j\beta r)}{r} \int_{v_1}^{v_2}\int_{u_1}^{u_2} u^{i-1} v^{j-1} \exp(j\beta \boldsymbol{r}' \,.\, \boldsymbol{i}_R)\, \mathrm{d}u\, \mathrm{d}v \quad (5.40)$$

We next obtain the far-zone approximation for the magnetic vector potential by substituting this expression for P_{ij} into eqn. 5.35, and then use eqns. 5.16 and 5.17 to calculate the far-zone electric-field vector and eqn. 5.20 to find the far-zone magnetic-field vector.

Note that all the integrals in the above expressions belong to only three types of relatively simple integral. This greatly facilitates the evaluation of potentials and field vectors.

5.3 Electromagnetic field of currents along generalised wires

5.3.1 Potentials and field vectors of currents along generalised wires

Recall that the u and v co-ordinates adopted for generalised wires are mutually perpendicular (see Figure 2.8), i.e. the angle α_{uv} between co-ordinate lines is $\pi/2$. Therefore, for generalised wires, $\sin\alpha_{uv} = 1$. Recall also the first part of eqn. 3.53, which we repeat here for convenience:

$$\mathrm{d}I_u = J_{su}\, \mathrm{d}l_v \sin\alpha_{uv} \quad (5.41)$$

Noting that $\mathrm{d}l_v = e_v\, \mathrm{d}v$, e_v being the Lamé coefficient, the u component of the surface-current-density vector is obtained as

$$J_{su} = \frac{1}{e_v}\frac{\partial I_u}{\partial v} \tag{5.42}$$

Without loss of generality, for any generalised wire we can always adopt the limits for the v co-ordinate in the form $v_1 = -v_2$. With this convention, and since we assume that the current along a generalised wire does not have a circumferential variation, we also have

$$J_{su} = \frac{1}{e_v}\frac{I_u}{v_2 - v_1} = \frac{1}{e_v}\frac{I(u)}{2v_2} \tag{5.43}$$

$I(u) = I_u$ representing the current intensity along the wire at the co-ordinate u. Combining the last two equations we thus show that, for generalised wires,

$$\frac{\partial I_u}{\partial v} = \frac{I(u)}{2v_2} \tag{5.44}$$

It is convenient to choose the origin of the v co-ordinates for any u in the following way. Note that the local v co-ordinate line has been chosen to be circular. Imagine the plane containing the field point and representing a plane of symmetry of that circle. The intersection of this plane and the circle will always be chosen as the origin of the v co-ordinates corresponding to the u co-ordinate considered. Choosing the origin of the v co-ordinates in this way and bearing in mind the last expression for $\partial I_u/\partial v$ and the convention $v_1 = -v_2$, the scalar and vector potentials in eqns. 5.21 and 5.22 become

$$V = \frac{j}{\omega\epsilon_0}\frac{1}{v_2}\int_{u_1}^{u_2}\frac{\mathrm{d}I(u)}{\mathrm{d}u}\int_0^{v_2} g(R)\,\mathrm{d}v\,\mathrm{d}u \tag{5.45}$$

$$\boldsymbol{A} = \mu_0\frac{1}{2v_2}\int_{u_1}^{u_2} I(u)\int_0^{v_2}\left\{\frac{\partial \boldsymbol{r}'(u,-v)}{\partial u} + \frac{\partial \boldsymbol{r}'(u,v)}{\partial u}\right\} g(R)\,\mathrm{d}v\,\mathrm{d}u \tag{5.46}$$

In a similar manner, for generalised wires the expressions for the electric-field vector and the magnetic-field vector in eqns. 5.23 and 5.24 take the form

$$\begin{aligned}\boldsymbol{E} = -j\omega\mu_0\frac{1}{2v_2}\Bigg[&\int_{u_1}^{u_2} I(u)\int_0^{v_2}\left\{\frac{\partial \boldsymbol{r}'(u,-v)}{\partial u} + \frac{\partial \boldsymbol{r}'(u,v)}{\partial u}\right\} g(R)\,\mathrm{d}v\,\mathrm{d}u\\ &+\frac{1}{\beta^2}\int_{u_1}^{u_2}\frac{\mathrm{d}I(u)}{\mathrm{d}u}\int_0^{v_2}\{\boldsymbol{i}_R(u,-v)+\boldsymbol{i}_R(u,v)\}\frac{\mathrm{d}g(R)}{\mathrm{d}R}\,\mathrm{d}v\,\mathrm{d}u\Bigg]\end{aligned} \tag{5.47}$$

and

$$\begin{aligned}\boldsymbol{H} = \frac{1}{2v_2}\int_{u_1}^{u_2} I(u)\int_0^{v_2}\Bigg\{&\boldsymbol{i}_R(u,-v)\times\frac{\partial \boldsymbol{r}'(u,-v)}{\partial u}\\ &+\boldsymbol{i}_R(u,v)\times\frac{\partial \boldsymbol{r}'(u,v)}{\partial u}\Bigg\}\frac{\mathrm{d}g(R)}{\mathrm{d}R}\,\mathrm{d}v\,\mathrm{d}u\end{aligned} \tag{5.48}$$

The far-zone field vectors are obtained in the same way as before. We first substitute in eqn. 5.46 $g(R)$ with its approximate form in eqn. 5.14, to obtain

$$A = \mu_0 \frac{1}{2v_2} \frac{\exp(-j\beta r)}{r} \int_{u_1}^{u_2} I(u) \times \int_0^{v_2} \left\{ \frac{\partial \boldsymbol{r}'(u, -v)}{\partial u} + \frac{\partial \boldsymbol{r}'(u, v)}{\partial u} \right\} \exp(j\beta \boldsymbol{r}' \cdot \boldsymbol{i}_R) \, \mathrm{d}v \, \mathrm{d}u \tag{5.49}$$

To find the far-zone electric-field vector we use eqns. 5.16 and 5.17, and to calculate the far-zone magnetic-field vector eqn. 5.20.

5.3.2 Reduced-kernel potentials and field vectors of currents along generalised wires

It is frequently of interest to analyse the system approximately by substituting the reduced kernel for the exact kernel. This simplifies numerical computations significantly with negligible loss in accuracy if thin wires are considered. The distance between the source point and the field point in this case is defined as (see Figure 5.2).

$$R_a = \sqrt{[\{\boldsymbol{r} - \boldsymbol{r}_a(u)\}^2 + a(u)^2]} \tag{5.50}$$

Since R_a is no longer a function of v, the expressions for the electric scalar potential and the magnetic vector potential in eqns. 5.45 and 5.46 reduce, as can be seen, to the following forms:

$$V = \frac{j}{\omega\epsilon_0} \int_{u_1}^{u_2} \frac{\mathrm{d}I(u)}{\mathrm{d}u} g(R_a) \, \mathrm{d}u \tag{5.51}$$

$$A = \mu_0 \int_{u_1}^{u_2} I(u) \frac{\mathrm{d}\boldsymbol{r}_a(u)}{\mathrm{d}u} g(R_a) \, \mathrm{d}u \tag{5.52}$$

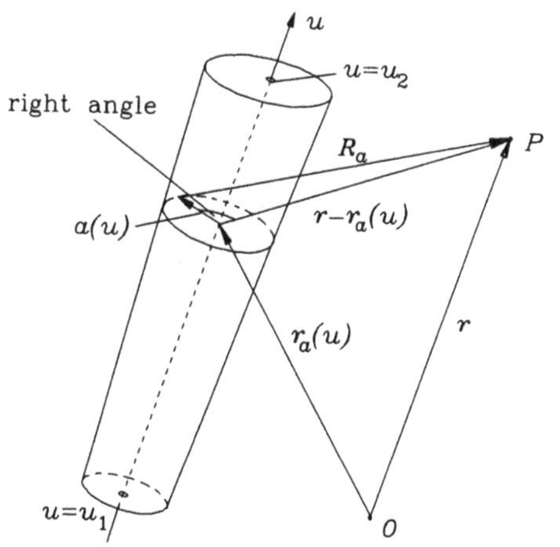

Figure 5.2 Generalised wire and the definition of the distance between source and field points entering the reduced kernel

To obtain the corresponding expressions for the electric- and magnetic-field vectors, note that the operators grad and curl can be introduced under the integral sign in eqns. 5.6 and 5.7, that they operate on $g(R_a)$ alone and that $\mathrm{grad}\{g(R_a)\}$ and curl $\{\boldsymbol{C}g(R_a)\}$, where $\boldsymbol{C}$ is a constant vector, can be written in the form

$$\mathrm{grad}\{g(R_a)\} = \frac{\mathrm{d}g(R_a)}{\mathrm{d}R_a}\,\mathrm{grad}(R_a) = \frac{\mathrm{d}g(R_a)}{\mathrm{d}R_a}\,\frac{\boldsymbol{r}-\boldsymbol{r}_a}{R_a} \tag{5.53}$$

and

$$\mathrm{curl}\{\boldsymbol{C}g(R_a)\} = \mathrm{grad}\{g(R_a)\} \times \boldsymbol{C} \tag{5.54}$$

With this in mind, it is simple to obtain the following expressions for the electric- and magnetic-field vectors corresponding to the reduced-kernel approximation:

$$\boldsymbol{E} = -j\omega\mu_0\left\{\int_{u_1}^{u_2} I(u)\,\frac{\mathrm{d}\boldsymbol{r}_a(u)}{\mathrm{d}u}\,g(R_a)\,\mathrm{d}u + \frac{1}{\beta^2}\int_{u_1}^{u_2}\frac{\mathrm{d}I(u)}{\mathrm{d}u}\,\mathrm{grad}(R_a)\,\frac{\mathrm{d}g(R_a)}{\mathrm{d}R_a}\,\mathrm{d}u\right\} \tag{5.55}$$

$$\boldsymbol{H} = \int_{u_1}^{u_2} I(u)\,\mathrm{grad}(R_a) \times \frac{\mathrm{d}\boldsymbol{r}_a(u)}{\mathrm{d}u}\,\frac{\mathrm{d}g(R_a)}{\mathrm{d}R_a}\,\mathrm{d}u \tag{5.56}$$

Substituting in the expressions for the potentials in eqns. 5.51 and 5.52 the expansion for current in eqn. 3.1, the expressions for the potentials become

$$V = \sum_{i=1}^{n_u} a_i V_i \tag{5.57}$$

$$V_i = \frac{j}{\omega\epsilon_0}\int_{u_1}^{u_2}\frac{\mathrm{d}f_i(u)}{\mathrm{d}u}\,g(R_a)\,\mathrm{d}u \tag{5.58}$$

and

$$\boldsymbol{A} = \sum_{i=1}^{n_u} a_i \boldsymbol{A}_i \tag{5.59}$$

$$\boldsymbol{A}_i = \mu_0\int_{u_1}^{u_2} f_i(u)\,\frac{\mathrm{d}\boldsymbol{r}_a(u)}{\mathrm{d}u}\,g(R_a)\,\mathrm{d}u \tag{5.60}$$

The field vectors are similarly obtained as

$$\boldsymbol{E} = \sum_{i=1}^{n_u} a_i \boldsymbol{E}_i \tag{5.61}$$

$$\boldsymbol{E}_i = -j\omega\boldsymbol{A}_i - \frac{j\omega\mu_0}{\beta^2}\int_{u_1}^{u_2}\frac{\mathrm{d}f_i(u)}{\mathrm{d}u}\,\mathrm{grad}(R_a)\,\frac{\mathrm{d}g(R_a)}{\mathrm{d}R_a}\,\mathrm{d}u \tag{5.62}$$

and

$$\boldsymbol{H} = \sum_{i=1}^{n_u} a_i \boldsymbol{H}_i \tag{5.63}$$

$$\boldsymbol{H}_i = \int_{u_1}^{u_2} f_i(u)\,\mathrm{grad}(R_a) \times \frac{\mathrm{d}\boldsymbol{r}_a(u)}{\mathrm{d}u} \frac{\mathrm{d}g(R_a)}{\mathrm{d}R_a}\,\mathrm{d}u \tag{5.64}$$

The far-zone field vectors are obtained in the same manner as before. By substituting $g(R_a)$, which can be obtained as $g(R)$ in eqn. 5.14, into eqn. 5.60, we first find the expression for the far-zone magnetic vector potential:

$$\boldsymbol{A}_i = \mu_0 \frac{\exp(-j\beta r)}{r} \int_{u_1}^{u_2} f_i(u) \frac{\mathrm{d}\boldsymbol{r}_a(u)}{\mathrm{d}u} \exp(j\beta \boldsymbol{r}_a \,.\, \boldsymbol{i}_R)\,\mathrm{d}u \tag{5.65}$$

In this expression, $\boldsymbol{r}$ is the position vector of the far-field point, $\boldsymbol{r}_a = \boldsymbol{r}_a(u)$ and $\boldsymbol{i}_R$ is the unit vector directed towards the far-field point. The far-zone electric-field vector is obtained from eqns. 5.16 and 5.17, and the far-zone magnetic-field vector from eqn. 5.20.

Note that the above expressions are not simplified for cylindrical wire segments, except that for the distance between the source and field point (eqn. 5.50) we have that $a(u)$ for cylindrical segments is a constant. This, however, is not a real simplification, because in both cases the expression under the square root is a square function of u.

5.3.3 Reduced-kernel potentials and field vectors of polynomial distribution of currents along generalised wires

As already pointed out, the polynomial approximation of currents along generalised wires appears to be very convenient both from analytical and numerical points of view. The complete expressions for the reduced-kernel potentials and field vectors with the approximation of (partial) currents in the form of power functions $[f_i(u) = u^i,\ i = 0, 1, 2, \ldots]$ are therefore derived separately.

If we take into account eqn. 2.20, the potentials and field vectors due to a power-function distribution of current along generalised wires become

$$V_i = \frac{j}{\omega\epsilon_0}(i-1)P_{i-1} \tag{5.66}$$

$$\boldsymbol{A}_i = \mu_0 \boldsymbol{r}_u P_i \tag{5.67}$$

$$\boldsymbol{E}_i = -j\omega\mu_0 \left[\boldsymbol{r}_u P_i + \frac{1}{\beta^2}(i-1)\{(\boldsymbol{r} - \boldsymbol{r}_c)Q_{i-1} - \boldsymbol{r}_u Q_i\}\right] \tag{5.68}$$

and

$$\boldsymbol{H}_i = (\boldsymbol{r} - \boldsymbol{r}_c) \times \boldsymbol{r}_u Q_i \tag{5.69}$$

where

$$\boldsymbol{r}_c = \boldsymbol{r}_1 - u_1\boldsymbol{r}_u \tag{5.70}$$

$$\boldsymbol{r}_u = \frac{\boldsymbol{r}_2 - \boldsymbol{r}_1}{u_2 - u_1} \tag{5.71}$$

$$P_i = \int_{u_1}^{u_2} u^{i-1} g(R_a)\,\mathrm{d}u \tag{5.72}$$

and

$$Q_i = \int_{u_1}^{u_2} u^{i-1} \frac{1}{R_a}\frac{\mathrm{d}g(R_a)}{\mathrm{d}R_a}\,\mathrm{d}u \tag{5.73}$$

In these equations, $\boldsymbol{r}_1$ and $\boldsymbol{r}_2$ are the position vectors of the starting and end points of the generalised wire axis (see Figure 5.2).

Finally, the corresponding expressions valid for the far zone are obtained if $g(R_a)$ in P_i is first replaced by its approximate expression, to obtain

$$P_i = \frac{\exp(-j\beta r)}{r}\int_{u_1}^{u_2} u^{i-1}\exp(j\beta\boldsymbol{r}_a\,.\,\boldsymbol{i}_R)\,\mathrm{d}u \tag{5.74}$$

As for the reduced kernel and arbitrary basis function, in this expression $\boldsymbol{r}$ is the position vector of the far-field point, $\boldsymbol{r}_a = \boldsymbol{r}_a(u)$ and $\boldsymbol{i}_R$ is the unit vector directed towards the far-field point. Substituting this expression for P_i into eqn. 5.67 we find the far-zone magnetic vector potential. We next obtain the far-zone electric-field vector from eqns. 5.16 and 5.17, and the far-zone magnetic-field vector from eqn. 5.20.

Note that the expressions in eqns. 5.66–5.69 have formally the same form as those in eqns. 5.34–5.37, if in the latter we set $\boldsymbol{r}_v = 0$ and $\boldsymbol{r}_{uv} = 0$, and if P_{ij} and Q_{ij} are replaced by P_i and Q_i, respectively. Consequently, the difference in calculating the potentials and field vectors of polynomial current distributions along truncated cones and over bilinear surfaces differs formally only in the use of different expressions for the P and Q integrals. Useful hints on the evaluation of the P and Q integrals can be found in the Appendices.

5.4 Conclusions

In this Chapter the missing link has been prepared for the derivation of the integral equation for the distribution of surface currents over generalised quadrilaterals and bilinear surfaces, as well as of currents along generalised wires and truncated cones. The general expressions for the potentials and the field vectors due to partial currents in the form of arbitrary functions have been derived, and partial power-function distributions of current have been considered as basis functions of polynomial approximation of currents. The expressions for far-zone field vectors have been derived in all instances.

Chapter 6

Solution of equations for current distribution

6.1 Introduction

In the preceding Chapters methods have been explained for approximation of the geometry of surfaces or wires and of current distributions over these surfaces or along these wires. Some approximations of the actual excitation of the structures have been proposed which permit relatively simple treatment of real excitation regions. The general expressions for the potentials and the electric- and magnetic-field vectors have also been derived.

Determination of the unknown coefficients in the expression for current distribution therefore represents the only remaining step in the analysis of metallic antennas and scatterers. As already mentioned, this can be achieved by solving numerically any of a number of known integral or integro-differential equations that can be formulated for the unknown current distribution, some of which were outlined in Section 1.3. This Chapter is devoted to discussing these equations in more detail, to justifying the adoption of one of them, and to explaining the reasons for adopting a specific procedure (the Galerkin method) for solving this equation for the unknown current-distribution coefficients. Finally, explicit expressions for the elements of the so-called impedance matrix in the Galerkin method are derived.

6.2 Comments on integral equations for current distribution

6.2.1 *Introduction*

As mentioned in Section 1.3, the basic integral equations for solving electromagnetic-field problems are the electric-field integral equation (EFIE) and the magnetic-field integral equation (MFIE). There are some other equations, however, which have been formulated and used in numerical analysis of metallic antennas and scatterers. This Section is devoted to a brief derivation, discussion and comparison of all these equations.

6.2.2 *Comments on basic field integral equations*

We know from Section 1.3 that the electric-field integral equation is obtained from the boundary condition for the tangential component of the electric-field vector on the surface of a perfect conductor, i.e.

$$\boldsymbol{n} \times \boldsymbol{E} + \boldsymbol{n} \times \boldsymbol{E}_i = 0 \quad \text{on} \quad S \tag{6.1}$$

where $\boldsymbol{E}$ stands for the electric-field vector of induced currents and charges, $\boldsymbol{E}_i$ for the impressed electric-field vector and S is the surface of the system (which may be composed of several subsurfaces). The general form of the electric-field integral equation is obtained if $\boldsymbol{E}$ in eqn. 6.1 is substituted by the expresson in eqn. 5.11, which we repeat here for convenience:

$$\boldsymbol{E} = -j\omega\mu_0 \left\{ \int_S \boldsymbol{J}_s g(R)\,\mathrm{d}S + \frac{1}{\beta^2} \int_S \mathrm{div}_s \boldsymbol{J}_s \boldsymbol{i}_R \frac{\mathrm{d}g(R)}{\mathrm{d}R}\,\mathrm{d}S \right\}$$

and $\boldsymbol{E}_i$ by the appropriate expression for the impressed field from Chapter 4.

It can be shown that the solution of this equation is not unique if there is a solution of the wave equation inside S which satisfies the same boundary conditions, i.e. if there is a solution of the equation $\Delta \boldsymbol{E} - \omega^2 \epsilon\mu \boldsymbol{E} = 0$ inside S which satisfies the condition $\boldsymbol{n} \times \boldsymbol{E} = 0$ on S (see, for example, Reference 74). (Recall that we consider S to be the surface defining the conducting bodies, that we are interested in the field outside conducting bodies, i.e. outside S, and that inside the conducting bodies the field is zero.) Obviously, this is the solution for the field inside a cavity of the shape defined by the surface S, made in a perfect electric conductor. In the general case the solution of the wave equation is not unique, i.e. there may be several modes at specific resonant frequencies, with different field distributions inside the cavity and different surface-current distributions over the cavity wall. Note that, theoretically, the currents on any closed cavity wall do not produce the field outside the cavity. Therefore the field outside S is unique, but not the currents on S, that may be a sum of currents producing the field outside S only and those producing the field inside S only.

However, in a numerical solution quite large errors in the outside field might be obtained. This can be explained briefly as follows. Note that the inside and outside fields are numerically coupled (although theoretically they are not). Note also that the specified boundary conditions on S cannot fix, for example, the amplitude of the inside field at a resonant frequency. This indeterminacy is transferred to the solution of the outside field. Mathematically speaking, in transforming the system of integral equations into a system of linear algebraic equations, ill-conditioned matrices are obtained in the vicinity of any resonant frequency. Hence, the EFIE cannot be used in the usual way for an efficient solution of either the resonant-body problem (to find the field outside the body), or that of an outside field coupled by means of an actual connection to a cavity, if the frequency is close to any of the cavity resonant frequencies.

Sarkar and Rao [88] proposed a minimum-norm solution for this instability of the EFIE at internal resonances. The procedure is an extension of the Murray method for linear-operator equations [89]. However, the solution of the EFIE by this method is still not exact for the currents, but only for the field outside the body [88].

Let us now make a few comments on the magnetic-field integral equation. We have seen in Section 1.3 that the magnetic-field integral equation is based on the boundary condition for the tangential component of the magnetic field vector just outside the surface of a perfect conductor. As explained in Section 5.2.1, if the boundary condition is used for formulating an equation adopted as the basis for numerical analysis of the field, it must be written in the form

$$\boldsymbol{n} \times (\boldsymbol{H} + 1/2\boldsymbol{J}_s \times \boldsymbol{n}) + \boldsymbol{n} \times \boldsymbol{H}_i = \boldsymbol{J}_s \tag{6.2}$$

In this equation, $\boldsymbol{H}$ is the magnetic-field vector given in eqn. 5.12, $\boldsymbol{H}_i$ is the impressed magnetic-field vector on S, and $\boldsymbol{J}_s$ is the local surface-current-density vector. The reason why the term $1/2\boldsymbol{J}_s \times \boldsymbol{n}$ must be added to $\boldsymbol{H}$ in formulating the boundary condition when numerical methods are used for field analysis was explained in comments following eqn. 5.12.

Noting that $\boldsymbol{n} \times (1/2\boldsymbol{J}_s \times \boldsymbol{n}) = 1/2\boldsymbol{J}_s$, eqn. 6.2 simplifies to

$$\boldsymbol{n} \times \boldsymbol{H} + \boldsymbol{n} \times \boldsymbol{H}_i = 1/2\boldsymbol{J}_s \tag{6.3}$$

If in this equation $\boldsymbol{H}$ is replaced by the expression in eqn. 5.12, which we repeat here for completeness,

$$\boldsymbol{H} = \int_S \frac{\mathrm{d}g(R)}{\mathrm{d}R}\, \boldsymbol{i}_R \times \boldsymbol{J}_s \,\mathrm{d}S$$

and $\boldsymbol{H}_i$ by the appropriate expression for the impressed field from Chapter 4, the magnetic-field-integral equation results.

Note that the MFIE as formulated in eqn. 6.3 requires that the field on the other (inner) side of the surface be zero. Therefore, it cannot be used for the analysis of infinitely thin open structures. For the same reason it is quite difficult to use the MFIE for the analysis of open thin metallic sheets. In the analysis of closed bodies at resonant frequencies it encounters basically the same difficulties as the EFIE. Moreover, in contrast to the EFIE, this equation does not yield a unique solution for the field outside the body at internal resonant frequencies of the body (see, for example, Reference 74).

6.2.3 Comments on augmented field integral equations

One possible way of obtaining a valid solution for closed bodies at internal resonant frequencies is as follows. The basic system of linear equations, obtained by starting from either EFIE or MFIE, can be extended by equations aimed at eliminating the resonant mode. An overdetermined system of equations is thus obtained which can be solved only by least-squares techniques, with the associated problem of the weights of individual equations of the overdetermined system.

For example, matching points can be adopted not only on the surface of the body, but also inside the body, at places where the resonant mode is known to be nonzero [68]. This obviously makes the choice of these matching points somewhat difficult. This method is sometimes termed the 'augmented-boundary-condition method' (ABC method) [90]. However, it belongs essentially to the same group as the following method.

Some authors (e.g. Yaghjian [91]) add to the EFIE or MFIE the equation for the normal component of the electric- (magnetic-) field vector just outside the surface of a perfect conductor, to obtain

$$\left.\begin{aligned} \boldsymbol{E} + \boldsymbol{E}_i &= \rho_s/2\epsilon \\ \text{or} \qquad & \\ \boldsymbol{H} + \boldsymbol{H}_i &= 1/2\boldsymbol{J}_s \times \boldsymbol{n} \end{aligned}\right\} \tag{6.4}$$

respectively. The first equation is usually termed the 'augmented electric-field integral equation' (AEFIE), and the second is known as the 'augmented magnetic field integral equation' (AMFIE). (Note that $\boldsymbol{E}$ in the first equation is given by the expression in eqn. 5.11, and $\boldsymbol{H}$ in the second equation by the expression in eqn. 5.12—see comments following those equations.) Explicit forms of these integral equations are obtained in the same way as those of the EFIE and MFIE.

When solving these equations by the method of moments, an overdetermined system of linear algebraic equations results with, theoretically, 50% more equations than the number of unknowns. For example, at all matching points two equations are obtained corresponding to the standard EFIE or MFIE, and one resulting from the boundary condition for the normal component of the electric-field vector or magnetic-field vector, respectively. (Actually, Yaghjian *et al.* added to every testing function for the tangential component a testing function for the normal component, so that the number of equations in their approach was twice the number of unknowns.)

6.2.4 Comments on combined field integral equation

To obtain a solution of an electromagnetic problem, it is also possible to start from a simultaneous system of equations resulting from both the EFIE and the MFIE. Of course, this yields an overdetermined system of equations which has a solution only in the least-squares sense. However, a unique solution can be obtained if the equations resulting from the EFIE and the MFIE are added together (with appropriate weights). The resulting equation yields an ordinary (not overdetermined) system of linear equations. This combination is termed the 'combined field integral equation' (CFIE). It was originally proposed by Oshiro [39].

Apparently most authors adopt precisely the CFIE for the analysis of metallic antennas and scatterers. As explained, it is of the form

$$\boldsymbol{n} \times \boldsymbol{H} + \frac{\alpha}{\mathcal{Z}} \boldsymbol{E}_{tang} + \boldsymbol{n} \times \boldsymbol{H}_i + \frac{\alpha}{\mathcal{Z}} \boldsymbol{E}_{i\,tang} = \boldsymbol{J}_s/2 \tag{6.5}$$

where α is arbitrary constant and $\mathcal{Z}$ is the intrinsic impedance of the medium. The explicit form of the CFIE is obtained if in eqn. 6.5 $\boldsymbol{H}$ is replaced by the expression in eqn. 5.12, $\boldsymbol{E}$ by the expression in eqn. 5.11, and $\boldsymbol{H}_i$ and $\boldsymbol{E}_i$ by appropriate expressions for impressed fields from Chapter 4.

It can be shown that, for $\alpha > 0$, the solution of this equation is also unique near resonant frequencies (see, for example, Reference 74). This approach does not result in an overdetermined system of algebraic equations, as does that based on AEFIE or AMFIE. However, the problem with eqn. 6.5 is the choice of the weighting coefficient α. Essentially, it is used to compare the electric-field vector with the magnetic-field vector, i.e. two different physical quantities, which makes the comparison quite arbitrary. Numerical results indicated that α should be chosen to be close to 1.

6.2.5 Comments on combined source integral equation

Finally, let us mention the so-called 'combined source integral equation' (CSIE, [92]). It solves the problem of instability of the results near internal resonant frequencies by introducing magnetic currents coupled with electric currents as sources of the induced electromagnetic field. This method yields the exact field outside the body, but does not give the current distribution over its surface.

6.2.6 Conclusions

Based on the above brief review of the existing integral equations for solving the problem of metallic antennas and scatterers, the following general conclusions concerning shortcomings of different equations can be drawn:

- The basic integral equations (EFIE and MFIE) are the simplest, but they fail at internal resonant frequencies which a metallic body might possess if it were hollow and infinitely thin-walled. MFIE cannot be used for the analysis of infinitely thin open structures.
- The combined source integral equation and the minimum-norm solution of the EFIE do not yield accurate current distributions, although they do yield accurate field vectors outside the body.
- All augmented equations result in an overdetermined system of equations, with the associated shortcomings.

Thus, it seems reasonable to conclude that the EFIE is probably optimal for open bodies (metallic sheets) and thin wires, and the CFIE for arbitary closed metallic bodies.

6.3 Comments on test procedures for numerical solution of integral equations for current distribution

6.3.1 Introduction

All the equations considered in Section 6.2 are so-called linear-operator equations, which can be written in a common form

$$\mathrm{L}f = g \tag{6.6}$$

where L is a linear operator, g is the known excitation function and f is the function to be determined. As explained in Section 1.3, numerical solution of such equations is always possible by the method of moments, in the manner explained there. However, the efficiency of the method depends greatly on the test procedure adopted.

In Section 6.3.2 it is shown that, when solving the operator equation of the form in eqn. 6.6 numerically, the three best-known procedures (point-matching, Galerkin and least-squares) are equivalent to a type of weighted point-matching procedure. Based on this equivalence, different procedures are compared, with the aim of choosing that which is optimal for the analysis of metallic antennas and scatterers.

6.3.2 Weighted point-matching procedure

For simplicity, consider first a 1-dimensional problem, and assume that the inner product has been adopted in the form

$$\langle w_j, p\rangle = \int w_j(z)^* p(z)\,\mathrm{d}z \qquad j = 1, 2, \ldots, N \tag{6.7}$$

where $p = \mathrm{L}f_i$, $i = 1, 2, \ldots, N$, or $p = g$, the asterisk denotes the complex conjugate and z is the local co-ordinate (for example, the co-ordinate along a cylindrical dipole axis). (Note that p and w_j are vector functions in the case of field analysis.) It was mentioned in Chapter 1 that the most frequent choices for testing functions w_j are the Dirac delta functions (resulting in the point-matching procedure), the chosen basis functions f_i (resulting in the Galerkin method), and functions of the form $w_j = \mathrm{L}f_j$ (resulting in the least-squares procedure).

The integral in the inner product in eqn 6.7 can only be evaluated numerically. In that case, however, instead of the inner product in eqn. 6.7 we in fact choose an inner product of the form

$$\langle w_j, p\rangle = \sum_{k=1}^{M} b_k w_j(z_k)^* p(z_k) \quad j = 1, 2, \ldots, N \tag{6.8}$$

where z_k are the arguments and b_k are the weighting coefficients of the formula adopted for numerical quadrature (e.g. the Gauss–Legendre quadrature formula). Consequently, the system of linear equations (eqn. 1.10) $\boldsymbol{FA} = \boldsymbol{G}$, where matrices $\boldsymbol{F}$, $\boldsymbol{A}$ and $\boldsymbol{G}$ were defined as $\boldsymbol{F} = [\langle w_j, \mathrm{L}f_i\rangle]_{N\times N}$, $\boldsymbol{A} = [A_i]_{N\times 1}$ (the matrix of unknown coefficients) and $\boldsymbol{G} = [\langle w_j, g\rangle]_{N\times 1}$, takes a different form. Now it reads

$$\left.\begin{aligned} (\boldsymbol{W}^*\boldsymbol{BF}_p)\boldsymbol{A} &= (\boldsymbol{W}^*\boldsymbol{BG}_p) \\ \text{or} \qquad & \\ (\boldsymbol{W}^*\boldsymbol{B})(\boldsymbol{F}_p\boldsymbol{A}) &= (\boldsymbol{W}^*\boldsymbol{B})\boldsymbol{G}_p \end{aligned}\right\} \tag{6.9}$$

where $\boldsymbol{W} = [w_j(z_k)]_{N\times M}$, $\boldsymbol{B} = \mathrm{diag}(b_1, \ldots, b_M)$, $\boldsymbol{F}_p = [\mathrm{L}f_i(z_k)]_{M\times N}$ and $\boldsymbol{G}_p = [g(z_k)]_{M\times 1}$, with $i, j = 1, 2, \ldots, N$ and $k = 1, 2, \ldots, M$. Note that the subscript i is the column number in matrices $\boldsymbol{F}$ and $\boldsymbol{F}_p$, that the subscript j is the row number in matrices $\boldsymbol{F}$ and $\boldsymbol{W}$, and that the subscript k is the column number in matrix $\boldsymbol{W}$ and the row number in matrix $\boldsymbol{F}_p$. Matrices $\boldsymbol{F}$ and $\boldsymbol{G}$ are labelled with the subscript p because they have the same form as in the point-matching procedure.

The two forms in eqn. 6.9 allow two interesting interpretations of the actual matrix equation we need to solve in a numerical solution of an operator equation by the method of moments. The first form is obviously the usual one, $\boldsymbol{FA} = \boldsymbol{G}$, except that the original matrices $\boldsymbol{F}$ and $\boldsymbol{G}$ are replaced by matrices $(\boldsymbol{W}^*\boldsymbol{BF}_p)$ and $(\boldsymbol{W}^*\boldsymbol{BG}_p)$, respectively.

Consider now the second form of the matrix equation. Note that the equation

$$\boldsymbol{F}_p\boldsymbol{A} = \boldsymbol{G}_p \qquad (6.10)$$

is precisely that which results from the point-matching procedure for the same problem, except that the number of matching points M in the general case is different from the number of unknowns N. (The matching points coincide with the arguments z_k in the chosen numerical-quadrature formula.) Obviously, this equation has a unique solution only if $M = N$, and a solution in the least-squares sense if $M > N$. We see that in eqn. 6.9 this system of equations is multiplied by the matrix $(\boldsymbol{W}^*\boldsymbol{B})$. Therefore the second form of the matrix equation can be considered as obtained by multiplying the (generally overdetermined) system of point-matching equations $(\boldsymbol{F}_p\boldsymbol{A}) = \boldsymbol{G}_p$ for the system considered by a weighting matrix $(\boldsymbol{W}^*\boldsymbol{B})$. Consequently, the method of moments, with the chosen inner product of the form in eqn. 6.7, should, whenever solved numerically (which in practice is always the case), be considered as a kind of weighted point-matching method [73].

It is not difficult to understand that eqns. 6.9 and 6.10 are also valid when the inner product is a surface integral encountered in the electromagnetic analysis of metallic surfaces. In that case, instead of the single-integration formula we have a product of two such formulas which performs the surface integration. Matching points are then the intersection points of the u and v co-ordinate lines. More precisely, the co-ordinates of the matching points are the v co-ordinates of the u co-ordinate lines, and the u co-ordinates of the v co-ordinate lines, corresponding to the arguments of the numerical quadrature formula.

From the above explanations it is obvious that, for any test procedure, the number of integration points M on the segment of integration implied by the inner product in eqn. 6.8 cannot be less than the number N of unknown coefficients, i.e. it must be $M \geqslant N$. If this condition is not satisfied, the point-matching system of equations $(\boldsymbol{F}_p\boldsymbol{A}) = \boldsymbol{G}_p$ does not have a solution, so that the actual system of equations (eqn. (6.9)) does not have a solution either. It can also be concluded that, for plates, the number of integration points per co-ordinate must not be less than the number of unknowns in that co-ordinate. If $M = N$, the weighted point-matching method yields the same result as the ordinary point-matching method [73].

The number M of integration points may be reduced by appropriate choice of the numerical quadrature formula. In the general case, for a given operator it is not possible to determine the optimum quadrature formula in advance, but it has to be found by numerical experiments. For example, in the analysis of wire and plate structures it is found that this integration is much more efficient if one uses the Gauss–Legendre quadrature formula instead of a simpler one (e.g. the midpoint rule).

6.3.3 Comments on the point-matching procedure

As already explained, the point-matching procedure is obtained if Dirac delta functions are adopted as weighting functions. In that case, the matrix $\boldsymbol{W}$ in eqn. 6.9 becomes an $N \times N$ unitary matrix, and $\boldsymbol{B}$ becomes a $N \times N$ unitary diagonal matrix.

Point matching is numerically the simplest, but it has a shortcoming that no general solution is available for the optimum choice of matching points [70, 72]. The matching points are usually chosen to be equidistant along a wire segment, or along a co-ordinate of a plate. However, the Galerkin method in many cases yields more accurate results than the point-matching method. On the basis of the above explanations, it can be inferred that the efficiency of the point-matching method in such cases could be improved if matching points were adopted according to the arguments of the Gauss–Legendre quadrature formula. Indeed, it has been found that in this manner the results obtained by point-matching become closer to those obtained by the Galerkin method.

For structures with interconnections, it appears that in the approach adopted in this book and with the chosen current expansion (singletons and doublets), it is not convenient to use the normal point-matching method. This is simple to understand from the following reasoning. With the normal point-matching method, a single matching point can be associated with any basis function in the expansion for currents. However, bearing in mind that the doublet terms are defined over two adjacent elements of the structure, positioning of the matching point somewhere along one or the other of these elements obviously cannot yield an accurate value for the doublet.

Therefore, instead of the normal point-matching method a modified point-matching method is proposed. In this modified method, the equations at matching points are written only for the singleton terms of the expansion. In so doing, the equations which correspond to specific singletons are written for the tangential component of the electric-field vector in the direction of the singletons. The value of the doublet term is determined by stipulating the condition that leakage of energy from the interconnection of two wires or a wire and a plate be equal to zero [33, 61], i.e. that energy leakage along the interconnection of two plates be zero. (The interconnection of two plates is divided into as many parts as there are doublet terms.) For this, it is sufficient to take one matching point per doublet on each side of the interconnection and to add up the two equations thus obtained.

Of course, this is not the only possibility. We can also require that the difference in the electric scalar potential of two interconnected wires at the interconnection be zero (or equal to the point-generator voltage, and/or to the voltage across the point impedance element, if such exist at the junction). For interconnection of plates, or of wires and plates, an analogous condition can be stipulated. For example, this condition was used in the program WIREZEUS for the analysis of wire antennas and scatterers [36, 37].

6.3.4 Comments on the least-squares procedure

The testing functions in the least-squares procedure are the functions obtained by the integral-equation operator acting on the basis functions. Consequently, in the least-squares method the matrix $\boldsymbol{W}$ is equal to $\boldsymbol{F}_p^t = [\mathrm{L}f_i(z_k)]_{N\times M}^t$, where the superscript t denotes transpose. On the other side, the matrix $\boldsymbol{B}$ when b_k, $k = 1, 2, \ldots, M$, are real (which they always are in the quadrature formulas that we need), can be represented formally in the form

$$\boldsymbol{B} = \boldsymbol{B}_{1/2}\boldsymbol{B}^*_{1/2} \tag{6.11}$$

where $\boldsymbol{B}_{1/2} = \mathrm{diag}(\sqrt{b_1}, \ldots, \sqrt{b_M})$. The second system of equations in eqn. 6.9 can be hence written as

$$(\boldsymbol{F}_p^{*t}\boldsymbol{B}_{1/2}^{*t})(\boldsymbol{B}_{1/2}\boldsymbol{F}_p)\boldsymbol{A} = (\boldsymbol{F}_p^{*t}\boldsymbol{B}_{1/2}^{*t})(\boldsymbol{B}_{1/2}\boldsymbol{G}_p) \tag{6.12}$$

or

$$(\boldsymbol{B}_{1/2}\boldsymbol{F}_p)^{*t}(\boldsymbol{B}_{1/2}\boldsymbol{F}_p)\boldsymbol{A} = (\boldsymbol{B}_{1/2}\boldsymbol{F}_p)^{*t}(\boldsymbol{B}_{1/2}\boldsymbol{G}_p) \tag{6.13}$$

Thus, numerical solution of the operator equation (eqn. 6.6) by the least-squares procedure and adoption of the inner product of the form in eqn. 6.7 actually implies the solution (in the least-squares sense) of the overdetermined system of equations (eqn. 6.10) multiplied by weighting coefficients $\sqrt{b_k}$, $k = 1, 2, \ldots, M$.

6.3.5 Comments of the Galerkin procedure

The testing functions for the Galerkin method are the same as the basis functions. Therefore the matrix $\boldsymbol{W}$ in eqn. 6.9 takes the form

$$\boldsymbol{W} = [f_j(z_k)]_{N\times M} \tag{6.14}$$

For generalised wires and nondimensional basis functions, the matrix

$$\boldsymbol{I} = \boldsymbol{A} \tag{6.15}$$

(the matrix of unknown coefficients) has the dimension of current intensity, the matrix

$$\boldsymbol{Z} = \boldsymbol{F} = \boldsymbol{W}^*\boldsymbol{B}\boldsymbol{F}_p \tag{6.16}$$

the dimension of impedance, and the matrix

$$\boldsymbol{V} = \boldsymbol{G} = \boldsymbol{W}^*\boldsymbol{B}\boldsymbol{G}_p \tag{6.17}$$

the dimension of voltage. For this reason the matrix $\boldsymbol{I} = \boldsymbol{A}$ is frequently termed the matrix of (generalised) currents in all cases, even when basis functions are not dimensionless; the matrix $\boldsymbol{V} = \boldsymbol{G}$ is known as the matrix of (generalised) voltages; and the matrix $\boldsymbol{Z} = \boldsymbol{F}$ is named the matrix of (generalised) impedances, or briefly the 'impedance matrix'. Obviously, a principal step in numerical implementation of the Galerkin method is evaluation of the impedance-matrix elements.

Compared with the point-matching procedure, the Galerkin method has a disadvantage of additional matrix multiplication. However, it does not require (arbitrary) choice of the positions of matching points, which is the principal shortcoming of the point-matching method. In addition less accurate solutions are obtained in some cases by point-matching than with the Galerkin method.

6.3.6 Comparison of computing time required by solution procedures

The principal part of the computing time in the analysis of metallic structures is divided between three main phases: that for computing the matrix $\boldsymbol{F}_p$, that needed for the multiplication of matrices $\boldsymbol{W}$ and $\boldsymbol{B}$ and $\boldsymbol{F}_p$, and that for the

Table 6.1 Principal terms in the expressions for the number of basic operations per phase in the analysis of metallic structures, for testing functions which have different definition bases

Definition domains of testing functions	*Evaluation of matrix **F***	*Multiplication of matrices **W**, **B** and $\boldsymbol{F_p}$*	*Solution of matrix equation*
Point	K_1N^2	–	$N^3/3$
Segment	K_1NM	MnN	$N^3/3$
Structure	K_1NM	MN^2	$N^3/3$

N = number of unknown coefficients
n = average number of unknowns per segment
M = number of matching points (integration points)
K_1 is estimated at about 100 for wires and somewhat larger for plates

solution of the system of linear equations. The number of operations needed by different solution procedures depends on the testing functions being defined at a point (point-matching method), along a segment (Galerkin method) or along the whole structure (least-squares method).

Table 6.1 gives the principal (leading) terms in the expressions for the number of basic operations per phase of the analysis for testing functions which have different definition domains. (By the basic operation we mean one addition and one multiplication.) In Table 6.1, K_1 stands for the number of operations necessary for computation of one element of the matrix $\boldsymbol{F}_p$, and n is the average number of unknowns per segment. For methods of analysis of thin-wire structures which are based on the electric-field integral equations, K_1 may roughly be estimated to be about 100 [28]. In the analysis of plates K_1 is somewhat larger.

Note that in the point-matching method the testing functions are always defined at a point, while in the least-squares method they are defined along all the segments of a structure. In contrast, the testing functions in the Galerkin method, as used in this book, are defined along one segment (singleton terms) and along two adjacent segments (doublet terms). Bearing this in mind, on the basis of Table 6.1 it is possible to determine the principal terms in the expressions for the total number of operations of the three solution procedures mentioned in special cases for small and for large N, as shown in Table 6.2. It is evident that, for determining the optimum solution procedure, the minimum values of the parameters M and N of interest are those which result in solutions of the desired accuracy. These values can only be determined by numerical experiments.

Table 6.2 Principal terms in the expressions for the number of basic operations in the analysis of metallic structures for different solving procedures in two special cases: for small and large N (number of unknown coefficients)

	Point-matching	Galerkin	Least-squares
Small N	K_1N^2	K_1MN	K_1MN
Large N	$N^3/3$	$N^3/3$	MN^2

M = number of matching points (integration points)
K_1 is estimated at about 100 for wires and at somewhat larger values for plates

6.3.7 Conclusions

By extensive numerical experiments, some of which are presented in Chapter 7, a presumably reliable estimate has been obtained of the efficiency of the solution procedures mentioned.

The results indicate that the method of least-squares is the most sensitive to accurate treatment of the end effects and that, when using relatively crude approximations for currents and charges near structure ends (wire ends and/or free plate sides), it is not possible to obtain reasonably accurate results even in the analysis of structures consisting solely of wires.

The point-matching method yields fairly acceptable results when used for the analysis of wire structures, on average perhaps slightly better if matching points are adopted according to the arguments of the Gauss–Legendre quadrature formula. However, in the analysis of plate structures the point-matching method frequently results in unstable solutions.

The Galerkin method does not exhibit any of the aforementioned shortcomings. It even requires a surprisingly low order of approximation. For example, for longer wires only three unknowns per wavelength, and for larger surfaces ten unknowns per square of the wavelength, appear to be sufficient. The number of integration points (as explained, actually the number of matching points in the weighted point-matching procedure) needs to be only a few times the number of unknowns. The Galerkin test procedure has therefore been adopted in this book as a conditionally optimum solution procedure in the analysis of metallic antennas and scatterers. Most of the numerical results presented in Chapters 7 and 8 have been obtained using this test procedure.

6.4 Impedance-matrix elements in the Galerkin method

6.4.1 General expressions

We have seen that in the Galerkin method the elements of the impedance matrix (which we shall frequently refer to briefly as the 'impedances') have the form $\langle f_k, \mathrm{L} f_l \rangle$. This inner product is calculated according to the surface form of eqn. 6.7, in which f_k is the kth testing function (the same as the kth basis function in the Galerkin method) and f_l is the lth basis function.

It is possible to choose the basis functions and the unknown coefficients in the expansion for surface currents to have various dimensions, but their product must have the dimension of surface-current density. For example, the basis functions may be dimensionless. However, for the derivations which follow in this Chapter it is convenient to assume that basis functions have the dimension of surface-current density and that a unit vector is attached to them in the local co-ordinate system (e.g. $\boldsymbol{i}_u$ for the u component of the surface current). It will therefore be clearer to write the functions f_k and f_l as $\boldsymbol{J}_{sk}$ and $\boldsymbol{J}_{sl}$, to stress that they have the dimension of the surface-current-density vector. They can be referred to either as basis functions k and l or as surface currents k and l. (Of course, the respective component of the total surface-current-density vector at a point of a surface is a sum of all basis functions defined over that surface.) In that case the expression $\mathrm{L} f_l$ represents the electric field of the surface current $\boldsymbol{J}_{sl}$, so it can be written as $\boldsymbol{E}_l$. From the above explanations, the impedance of these two basis functions (valid for the EFIE) is given by the expression

$$Z_{kl}^{EFIE} = \int_{S_k} \boldsymbol{J}_{sk} \cdot \boldsymbol{E}_l \, \mathrm{d}S_k \tag{6.18}$$

where S_k is the surface over which the kth basis function (surface current) is defined. Note that the dot product ensures that only the tangential component of $\boldsymbol{E}_l$ enters the expression. The corresponding part of the free-terms matrix is also given by eqn. 6.18, except that the electric-field vector $\boldsymbol{E}_l$ due to the lth basis function (surface current) should be replaced by the incident electric field $\boldsymbol{E}_i$.

If $\boldsymbol{E}_l$ is expressed in terms of the electric scalar potential and the magnetic vector potential, $\boldsymbol{E}_l = -\mathrm{grad}\, V_l - j\omega \boldsymbol{A}_l$, we can express the impedance of basis functions k and l in terms of surface charges, surface currents and potentials. We first have

$$Z_{kl}^{EFIE} = -\int_{S_k} \boldsymbol{J}_{sk} \cdot \mathrm{grad}\, V_l \, \mathrm{d}S_k - j\omega \int_{S_k} \boldsymbol{J}_{sk} \cdot \boldsymbol{A}_l \, \mathrm{d}S_k \tag{6.19}$$

Note that

$$\begin{aligned} \mathrm{div}_s(V_l \boldsymbol{J}_{sk}) &= V_l\, \mathrm{div}_s \boldsymbol{J}_{sk} + \boldsymbol{J}_{sk} \cdot \mathrm{grad}_s V_l \\ &= -j\omega \rho_{sk} V_l + \boldsymbol{J}_{sk} \cdot \mathrm{grad}\, V_l \end{aligned} \tag{6.20}$$

and (see, for example, Reference 86)

$$\int_{S_k} \mathrm{div}_s(V_l \boldsymbol{J}_{sk}) \, \mathrm{d}S_k = \int_{C_k} V_l \boldsymbol{J}_{sk} \cdot \mathrm{d}\boldsymbol{c}_k = 0 \tag{6.21}$$

where div_s stands for surface divergence and $\mathrm{d}\boldsymbol{c}_k = \boldsymbol{i}_k \, \mathrm{d}c_k$, $\boldsymbol{i}_k$ being unit vector perpendicular to the contour and in the local plane tangential to the contour. With this in mind, an alternative expression for the impedance of basis functions k and l (valid for the EFIE) is obtained:

$$Z_{kl}^{EFIE} = -j\omega \int_{S_k} (\rho_{sk} V_l + \boldsymbol{J}_{sk} \cdot \boldsymbol{A}_l) \, \mathrm{d}S_k \tag{6.22}$$

At first glance, it might seem that the expression in eqn. 6.18, being simpler, is more convenient than that in eqn. 6.22. However, the expression in eqn. 6.22 is more convenient for two reasons. First, the potentials can be computed easier than the field. Secondly, since the potentials thus obtained are less quasisingular than the field functions, (numerical) evaluation of the integrals in eqn. 6.22 is easier than that in eqn. 6.18.

Starting from either eqn. 6.18 or eqn. 6.22, it can easily be shown that the impedance of two basis functions k and l is symmetrical, i.e. that $Z_{kl}^{EFIE} = Z_{lk}^{EFIE}$. This permits faster evaluation of the impedance matrix and the solution of the matrix equation.

In particular, if the problem is formulated in terms of the CFIE, the impedance of basis functions k and l is

$$Z_{kl}^{CFIE} = \frac{\alpha}{Z} Z_{kl}^{EFIE} + \int_{S_k} \boldsymbol{J}_{sk} \cdot \left(\boldsymbol{n} \times \boldsymbol{H}_l - \frac{1}{2} \boldsymbol{J}_{sl} \right) \mathrm{d}S_k \tag{6.23}$$

The surface current $\boldsymbol{J}_{sl}$ is needed only on that part of the surface where it overlaps the surface current $\boldsymbol{J}_{sk}$. Note that $Z_{kl}^{CFIE} \neq Z_{lk}^{CFIE}$.

6.4.2 *Impedance of generalised quadrilaterals and generalised wires*

Assume first again that the problem is formulated in terms of the EFIE and let us determine the impedance of two basis functions defined over generalised quadrilaterals or generalised wires. If the expressions for $\boldsymbol{J}_s$ in eqns. 3.52, 3.59 and 3.60 are taken into account, after simple transformations and making use of the expressions in eqns. 2.7–2.10, eqn. 6.18 can be written as

$$Z_{kl}^{EFIE} = \int_{v_{1k}}^{v_{2k}} \int_{u_{1k}}^{u_{2k}} \left\{ \left(\frac{\mathrm{d}I_u}{\mathrm{d}v} \right) \frac{\mathrm{d}\boldsymbol{r}}{\mathrm{d}u} \right\}_k . \boldsymbol{E}_l \, \mathrm{d}u_k \, \mathrm{d}v_k \tag{6.24}$$

where $\boldsymbol{E}_l$ stands for the desired electric-field term $\boldsymbol{E}_{ij}$ in eqn. 5.30. In a similar manner the alternative expression in eqn. 6.22 can be transformed into

$$Z_{kl}^{EFIE} = \int_{v_{1k}}^{v_{2k}} \int_{u_{1k}}^{u_{2k}} \left[\left\{ \frac{\mathrm{d}}{\mathrm{d}u} \left(\frac{\mathrm{d}I_u}{\mathrm{d}v} \right) \right\}_k V_l - j\omega \left\{ \left(\frac{\mathrm{d}I_u}{\mathrm{d}v} \right) \frac{\mathrm{d}\boldsymbol{r}}{\mathrm{d}u} \right\}_k . \boldsymbol{A}_l \right] \mathrm{d}u_k \, \mathrm{d}v_k \tag{6.25}$$

with V_l being the desired V_{ij} given by eqn. 5.26, and $\boldsymbol{A}_l$ is the desired $\boldsymbol{A}_{ij}$ given by eqn. 5.28. Thus, simple expressions have been obtained, for evaluating the impedance of two basis functions, which do not contain the Lamé coefficients.

If the problem is formulated in terms of the CFIE, the impedance is obtained as

$$Z_{kl}^{CFIE} = \frac{\alpha}{Z} Z_{kl}^{EFIE} + \int_{v_{1k}}^{v_{2k}} \int_{u_{1k}}^{u_{2k}} \left\{ \left(\frac{\mathrm{d}I_u}{\mathrm{d}v} \right) \frac{\mathrm{d}\boldsymbol{r}_a}{\mathrm{d}u} \right\}_k . \left\{ \boldsymbol{n} \times \boldsymbol{H}_l - \frac{1}{2} \boldsymbol{J}_{sl} \right\} \mathrm{d}u_k \, \mathrm{d}v_k \tag{6.26}$$

These expressions are valid both for generalised quadrilaterals and generalised wires. For testing functions defined along generalised wires, however, evaluation of the impedance can further be simplified. We assume that the electric field due to basis functions is approximately constant along the circumference of the wire for which the testing function is defined, and that it is equal to the electric field along the wire axis. Bearing in mind that

$$\mathrm{d}S_k = \mathrm{d}l_{uk} \, \mathrm{d}l_{vk} \tag{6.27}$$

because $\sin \alpha_{uv} = 1$ for generalised wires, and that

$$\boldsymbol{J}_{sk} = J_{suk} \boldsymbol{i}_{uk} \tag{6.28}$$

(this is the same expression as in eqn. 3.52, only with subscript k added to all quantities), the impedance in eqn. 6.18 can be written in the form

$$Z_{kl}^{EFIE} = \int_{S_k} \boldsymbol{J}_{sk} . \boldsymbol{E}_l \, \mathrm{d}S_k \simeq \int_{u_{1k}}^{u_{2k}} J_{suk} \boldsymbol{E}_l . \left(\int_{v_{1k}}^{v_{2k}} \boldsymbol{i}_{uk} \, \mathrm{d}l_{vk} \right) \mathrm{d}l_{uk} \tag{6.29}$$

The expression in parentheses can be evaluated to obtain

$$\int_{v_{1k}}^{v_{2k}} \boldsymbol{i}_{uk} \, \mathrm{d}l_{vk} = 2\pi a_k \boldsymbol{i}_{zk} \cos \alpha_k \tag{6.30}$$

where α_k is the angle between the generalised wire axis and its generatrix, that, in general, is a function of the local co-ordinate u_k.

Noting that

$$I_k = J_{suk} 2\pi a_k \tag{6.31}$$

where I_k is a basis function defined along the generalised wire and having dimension of current intensity, and that

$$dz_k = dl_{uk} \cos \alpha_k \tag{6.32}$$

we obtain for the impedance expression

$$Z_{kl}^{EFIE} \simeq \int_{z_{1k}}^{z_{2k}} I_k \boldsymbol{i}_{zk} \cdot \boldsymbol{E}_l \, dz_k = \int_{z_{1k}}^{z_{2k}} I_k E_{lz} \, dz_k \tag{6.33}$$

In this equation, z_k is the local length co-ordinate along the generalised wire axis.

Note that the expressions for charge per unit length of the wire, i.e. per unit length of the wire z axis, read

$$Q'_{ku} = \frac{j}{\omega} \frac{dI_k}{dl_{uk}} = \rho_{sk} 2\pi a_k \tag{6.34}$$

and

$$Q'_{kz} = \frac{j}{\omega} \frac{dI_k}{dz_k} = Q'_{ku} \frac{dl_{uk}}{dz_k} \tag{6.35}$$

respectively. Bearing this in mind, the impedance in eqn. 6.22 can be written in the form

$$Z_{kl}^{EFIE} \simeq -j\omega \int_{z_{1k}}^{z_{2k}} (Q'_{kz} V_l + I_k A_{lz}) \, dz_k \tag{6.36}$$

It is interesting and useful to note that the expressions in eqns. 6.33 and 6.36, where single integrals only need to be evaluated, can be transformed into the form of the expressions in eqns. 6.24 and 6.25, which include dual integrals. This formal identity makes the actual evaluation of the impedance-matrix elements for combined wire and plate structures significantly easier, as will be explained in Section 6.4.3.

First, note that the expression for the current intensity I_k can be written in the form

$$I_k = (v_{2k} - v_{1k}) \left(\frac{dI_u}{dv} \right)_k = \left(\frac{dI_u}{dv} \right)_k \int_{v_{1k}}^{v_{2k}} dv_k \tag{6.37}$$

Substituting this into the expression for the impedance in eqn. 6.33, and noting that

$$\boldsymbol{i}_{zk} \, dz_k = \frac{d\boldsymbol{r}_{ak}}{du_k} \, du_k \tag{6.38}$$

the expression for the impedance in eqn. 6.33 can be written as

$$Z_{kl}^{EFIE} \simeq \int_{v_{1k}}^{v_{2k}} \int_{u_{1k}}^{u_{2k}} \left\{ \left(\frac{dI_u}{dv} \right) \frac{d\boldsymbol{r}_a}{du} \right\}_k \cdot \boldsymbol{E}_l \, du_k \, dv_k \tag{6.39}$$

This is exactly the same as eqn. 6.24, except that $\boldsymbol{r}$ (the position vector in the parametric description of the generalised quadrilateral) is replaced by $\boldsymbol{r}_a$ (the position vector in the parametric equation of the generalised wire axis), so that the integrand in the last integral is not a function of the v co-ordinate.

In a similar manner, starting from the expression

$$Q'_{kz} = \frac{j}{\omega}\frac{\mathrm{d}}{\mathrm{d}z_k}\left(\frac{\mathrm{d}I_u}{\mathrm{d}v}\right)_k (v_{2k} - v_{1k}) = \frac{j}{\omega}\frac{\mathrm{d}}{\mathrm{d}z_k}\left(\frac{\mathrm{d}I_u}{\mathrm{d}v}\right)_k \int_{v_{1k}}^{v_{2k}} \mathrm{d}v_k \tag{6.40}$$

the impedance in eqn. 6.36 can be transformed into

$$Z_{kl}^{EFIE} = \int_{v_{1k}}^{v_{2k}} \int_{u_{1k}}^{u_{2k}} \left[\left\{\frac{\mathrm{d}}{\mathrm{d}u}\left(\frac{\mathrm{d}I_u}{\mathrm{d}v}\right)\right\}_k V_l - j\omega\left\{\left(\frac{\mathrm{d}I_u}{\mathrm{d}v}\right)\frac{\mathrm{d}\boldsymbol{r}_a}{\mathrm{d}u}\right\}_k \cdot \boldsymbol{A}_l\right] \mathrm{d}u_k\, \mathrm{d}v_k \tag{6.41}$$

The formal similarity between this expression, with the testing functions along the wire, and that in eqn. 6.25, where the testing functions are over generalised quadrilaterals, is evident. Again $\boldsymbol{r}$ (the position vector in the parametric description of the generalised quadrilateral) is replaced by $\boldsymbol{r}_a$ (the position vector in the parametric equation of the generalised wire axis), so that the integrand in eqn. 6.41 is not a function of the v co-ordinate.

6.4.3 *Impedance of bilinear surfaces and truncated cones with polynomial approximation of current*

Consider the impedance of two basis functions defined over two bilinear surfaces, starting from the expression in eqn. 6.24. As already pointed out, in this book current expansions are used which automatically satisfy the current-continuity equation along bilinear surface edges and along interconnections of two or more bilinear surfaces. These expansions are obtained starting from the ordinary double expansions given in eqn. 3.60, defined over a single bilinear surface. It is evident that the impedances of basis functions satisfying the continuity equation along edges and interconnections can be obtained combining the impedances of the initial basis functions given in eqn. 3.60. Therefore we shall consider henceforth only impedances of the initial basis functions.

Let the kth basis function be defined by subscripts $i = i_m$ and $j = j_m$ in the expansion in eqn. 3.60 defined for the mth bilinear surface, and let the lth basis function be defined by subscripts $i = i_n$ and $j = j_n$ in the expansion in eqn. 3.60 defined for the nth bilinear surface. Also, let the functions $f_i(u)$ and $h_j(v)$ in the expansion (eqn. 3.60) be power functions. We next introduce into eqn. 6.24 the expressions for the potentials in eqns. 5.34 and 5.35. After simple transformations we finally obtain

$$\begin{aligned} Z_{i_m j_m i_n j_n}^{EFIE} = -j\omega\mu_0 \Bigg\{ & -\frac{1}{\beta^2}(i_m - 1)(i_n - 1) S_{i_m-1, j_m, i_n-1, j_n} \\ & + (\boldsymbol{r}_{um} \cdot \boldsymbol{r}_{un}) S_{i_m, j_m, i_n, j_n} + (\boldsymbol{r}_{um} \cdot \boldsymbol{r}_{uvn}) S_{i_m, j_m, i_n, j_n+1} \\ & + (\boldsymbol{r}_{uvm} \cdot \boldsymbol{r}_{un}) S_{i_m, j_m+1, i_n, j_n} + (\boldsymbol{r}_{uvm} \cdot \boldsymbol{r}_{uvn}) S_{i_m, j_m+1, i_n, j_n+1} \Bigg\} \end{aligned} \tag{6.42}$$

where

$$S_{i_m, j_m, i_n, j_n} = \int_{v_{1k}}^{v_{2k}} \int_{u_{1k}}^{u_{2k}} u^{i_m - 1} v^{j_m - 1} P_{i_n j_n} \, dv_m \tag{6.43}$$

and $P_{i_m j_m}$ are given in eqn. 5.38. Numerical experiments indicated both that for the evaluation of these integrals single precision suffices, and that the Gauss–Legendre quadrature formula is more efficient than the others, e.g. than the Newton–Cotes quadrature formulas or the midpoint rule.

The above expressions show that the evaluation of the impedance-matrix elements of two bilinear surfaces is reduced to a single class of integrals. Recall that the surface current over a bilinear surface is represented as a superposition of the u components of the surface-current-density vector of two overlapping bilinear surfaces. However, in practice only one of the components is the u-component, while the other is the v component. This means that, for two initial bilinear surfaces, there are three more classes of impedance in addition to that considered above, because each of the two current components of one bilinear surface interacts with both current components of the other. These three classes of impedance element are given by the above expression, however, except that for one surface (or both surfaces) the vector $\boldsymbol{r}_u$ must be replaced by the vector $\boldsymbol{r}_v$, and that the subscripts i and j exchange places. It is evident that, in evaluating any of the four impedances, we need to solve the same class of integrals. This means that, in evaluating the four impedances, these integrals need to be evaluated only once. This represents the main part of the computation, so that only a minor part, that of obtaining the impedances from these integrals, need be performed four times.

It was shown in Section 5.3.3 that the expressions for the potentials of truncated cones can formally be written in the same form as the expressions for the potentials of bilinear surfaces. (This is based on the fact that the equation of the cone axis is a special case of the bilinear surface equation: for cones it is only necessary to set $\boldsymbol{r}_v = 0$ and $\boldsymbol{r}_{uv} = 0$.) It was also shown (Section 6.4.2) that the general expressions for impedances of generalised quadrilaterals and generalised wires in terms of the potentials can be written in essentially the same form. With this in mind, it is not difficult to demonstrate that the above expressions for the impedance-matrix elements are valid for any of the four combinations of bilinear surface and truncated cone, with the sole difference that for cones $\boldsymbol{r}_v = 0$ and $\boldsymbol{r}_{uv} = 0$. As far as the integrals S_{i_m, j_m, i_n, j_n} are concerned, they are evaluated in the same way for testing functions along wires and for those over surfaces, with the only difference that for wires $j_k = 0$ and the integrand is not a function of the v co-ordinate. In other words, a single point is sufficient for numerical integration along the v co-ordinate. This means that the difference in the evaluation of the impedances for these four combinations of bilinear surfaces and truncated cones is only in the evaluation of the integrals $P_{i_n j_n}$. For these integrals we have two evaluation procedures: one for bilinear surfaces, and one for truncated cones (see Appendices).

6.5 Conclusions

The first part of this Chapter was devoted to the discussion of possible integral equations and their solution from the point of view of the analysis of metallic antennas. It has been concluded that, apparently, the EFIE seems to be the optimum for open structures and thin wires, while the CFIE is probably the optimum for metallic bodies which cannot be approximated by thin sheets.

The next part of the Chapter was devoted to a comparison of different test procedures for numerical solution of integral equations for current distribution. The conclusion was arrived at that the Galerkin method seems to be the optimum one.

In the third part of the Chapter the general expressions were derived for the impedance of two basis functions defined for two arbitrary surface elements. In particular, the impedances were derived for bilinear surfaces and truncated cones with a polynomial approximation of current, which was adopted in this book as the most convenient current approximation.

Chapter 7

Numerical examples illustrating the choice of optimum elements of the method

7.1 Introduction

This and the next, final, Chapter are aimed at illustrating numerically the method for the analysis of metallic antennas and scatterers elaborated in the preceding Chapters. In this Chapter several groups of examples are presented which are intended to justify various choices made in the method adopted for the analysis. The first group consists of examples illustrating the choice of the method for modelling of the structure geometry of both metallic bodies and wires. The second group contains examples which illustrate the reasons that have led to the choice of the method for the approximation of currents over generalised quadrilaterals and along generalised wires. These are followed by a group of examples illustrating modelling of excitation. A final group of examples is intended to assist understanding of the reasons for the particular choice of the test procedure adopted in the book.

If not stated otherwise, the following assumptions apply to all the examples:

- Modelling of geometry in the examples is done by truncated cones and bilinear surfaces
- Modelling of junctions is based on the general localised junction model
- The current is approximated by basis functions which automatically satisfy the continuity equation at wire ends, plate edges and all interconnections, as well as the continuity of charge at wire ends
- In most cases EFIE is used to determine currents. Only for closed bodies is CFIE used instead
- The equation for current distribution is solved by the Galerkin method
- Wherever possible, symmetry is taken into account to expedite computations.

An important point should be made about the philosophy of the solutions presented in this and the next Chapter. The purpose of the examples that follow is, in the first place, to analyse the systems by the entire-domain approach. The Galerkin method, owing to the presence of another integration, appears to be quite insensitive to local rapid variations of the impressed field, such as occur in magnetic-current-frill excitation of thin wire antennas. The point-matching procedure, on the other hand, requires that in such cases a sufficient number of matching points be distributed in the region of the rapidly varying excitation field, and therefore needs partitioning of the segment, very high-order of approximation along the whole segment, or possibly a different distribution of matching points instead of the usual equidistant distribution.

The aim of this book, and of the examples which follow, is to apply the entire-domain, or almost-entire domain, philosophy in all the examples and in all the procedures (except for the results of other authors), because it is conceptually the simplest (it does not require any arbitrary partitioning of the object). The comparisons of different methods given in the following examples have to be understood in this context. It is certain that, for example, with the point-matching procedure one can obtain much better results and with far fewer unknowns, if the object, e.g. a wire antenna, is partitioned appropriately, than if it is not partitioned. Nevertheless, whenever possible all the procedures were tested using the entire-domain approach, to indicate the superiority of the Galerkin method in that case. It is very important to remember this, for otherwise the possibilities of the other procedures may be seriously misinterpreted.

Finally, although the proposed method in most cases yields accurate, or at least acceptable results with significantly fewer unknowns, in the following examples a degree of approximation was in some cases adopted which approximately corresponded to that of another theoretical method. This was done to demonstrate the greater accuracy of the proposed method, compared with the other method, if approximately the same degree of approximation is adopted.

7.2 Examples illustrating the choice of the method for modelling of geometry

The examples in this Section mainly illustrate some problems relating to modelling of geometry. Section 7.2.1 illustrates modelling of wire ends. Section 7.2.2 discusses segmentation of curved wires. Sections 7.2.3 and 7.2.4 illustrate modelling of curved surfaces. Finally, Sections 7.2.5–7.2.9 illustrate the modelling of wire-to-plate junctions.

7.2.1 Effect of modelling of ends of resonant-wire monopole antenna

It is well known that the treatment of wire ends influences the solution for current distribution along the wires. This effect is particularly pronounced for wires of approximately resonant lengths, or wires representing part of a resonant structure. A typical example is a relatively thick resonant vertical monopole antenna above a perfectly conducting ground plane base-driven by a TEM magnetic-current frill, the symmetrical equivalent of which is shown in Figure 7.1

The antenna was analysed using two geometrical models. In both cases the cylindrical part of the monopole remained the same. In the first case the flat end was not taken into account, while in the second it was modelled by a flat disc. The dipole current was approximated by basis functions which, among others, satisfied the quasistatic relation at the frill. The current distribution along the monopole (i.e. the dipole arm) was adopted as a polynomial of degree N, and

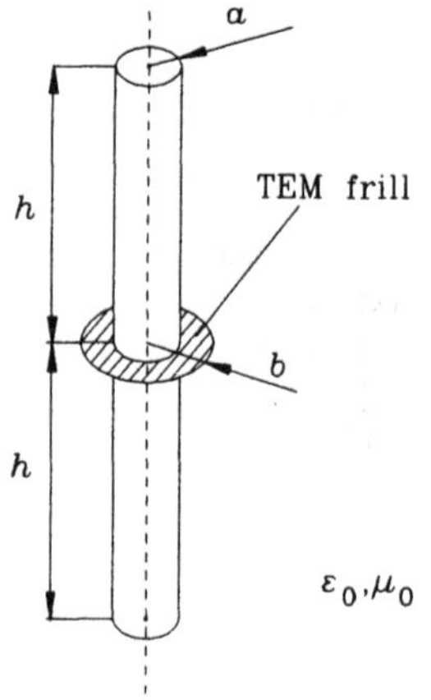

Figure 7.1 *Symmetrical equivalent of a vertical monopole antenna above a perfectly conducting ground plane base-driven by TEM magnetic-current frill*

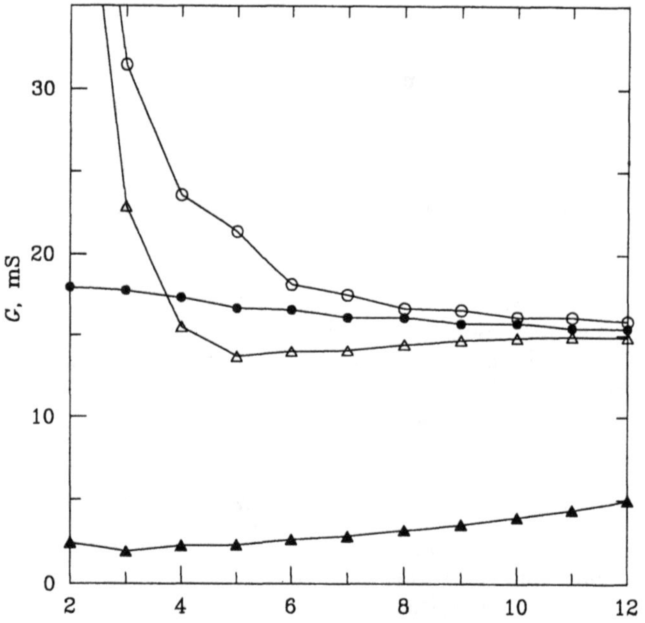

Figure 7.2 *Conductance of the monopole antenna the symmetrical equivalent of which is sketched in Figure 7.1, against the degree N of the polynomial approximation of current along the cylindrical dipole part*

The end effect is not taken into account
$h/a = 25$, $b/a = 2.3$ and $h = \lambda_0/4$
○ ○ ○ point-matching, equidistant matching points
△ △ △ point-matching, matching points at arguments of the Gauss–Legendre integration formula
● ● ● Galerkin
▲ ▲ ▲ least-squares

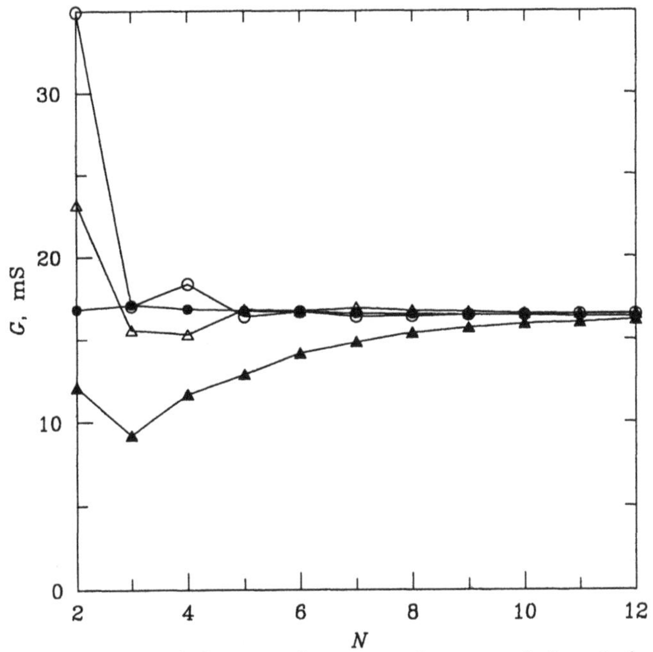

Figure 7.3 *Conductance of the monopole antenna the symmetrical equivalent of which is sketched in Figure 7.1 against the degree N of the polynomial approximation of current along the cylindrical dipole part*

The end effect is taken into account
$h/a = 25$, $b/a = 2.3$ and $h = \lambda_0/4$
○○○ point-matching, equidistant matching points
△△△ point-matching, matching points at arguments of the Gauss–Legendre integration formula
●●● Galerkin
▲▲▲ least-squares

along the radius of the flat end as a 2-degree polynomial. The solution of the EFIE was performed by four test procedures: (i) the point-matching procedure with equally spaced matching points, (ii) the point-matching procedure with matching points adopted at the arguments of the Gauss–Legendre quadrature formula, (iii) the Galerkin method, and (iv) the least-squares procedure.

Figures 7.2 and 7.3 show the conductance of a monopole with $h/a = 25$ and $b/a = 2.3$, assuming that the height of the monopole is exactly a quarter wavelength (in free space), plotted against the degree N of the polynomial. The results in Figure 7.2 were obtained without taking into account the end effect, while those in Figure 7.3 were obtained taking this effect into account. It is seen that accuracy, stability and convergence are significantly improved by taking the end effect into account.

It was found that, if the Galerkin method is used, taking into account the quasistatic relation at the frill somewhat slows down the convergence of the antenna admittance. This, however, significantly improves convergence if point-matching and least-squares procedures are used instead. (This is not very

Table 7.1 Theoretical and experimental results for admittance of a vertical monopole antenna with hemispherical end base-driven by TEM magnetic-current frill

	Type of wire end	Y (mS)
1	Equivalent flat end (num. result)	$18.60-j7.49$
2	Equivalent conical end (num. result)	$18.02-j7.65$
3	Hemispherical end (experimental result)	$17.84-j7.50$

$h=75$ mm, $a=2.1$ mm, $b=6.3$ mm and $h=\lambda_0/4$
Theoretical results were obtained by two approximate models of the wire end, as indicated, with $N=2$ unknowns along the cylindrical-antenna part

pronounced with relatively thick antennas, but with thin antennas, e.g. with $h/a = 100$ or greater, this becomes relevant.)

Table 7.1 shows the theoretical and experimental results for the admittance of a monopole antenna with a hemispherical end, with $h = 75$ mm, $a = 2.1$ mm, $b = 6.3$ mm and $h = \lambda_0/4$. The theoretical results were obtained by the Galerkin method, with two approximations of the hemispherical antenna end indicated in Figure 2.7*d*, and adopting $N = 2$ unknowns for the cylindrical-antenna part. As expected, more accurate results are obtained with the conical approximation of the spherical end than with the flat-disc approximation. Note that there is actually no difference in complexity in modelling the two types of end.

7.2.2 *Modelling of a ring scatterer by straight-wire segments*

Figure 7.4 shows a ring scatterer lying in the $y0z$ plane. The scatterer is modelled as (*a*) a single generalised wire, and (*b*) with $S = 6$ cylindrical segments. In the first case, entire-domain approximation was used for the whole ring (polynomial of degree six, with a total of six unknowns); in the second case, the subdomain approximation (first degree of the polynomial, resulting in a total of six unknowns for the entire ring) and an entire-domain approximation (the second degree of the polynomial, resulting in a total of 12 unknowns for the entire ring) were used.

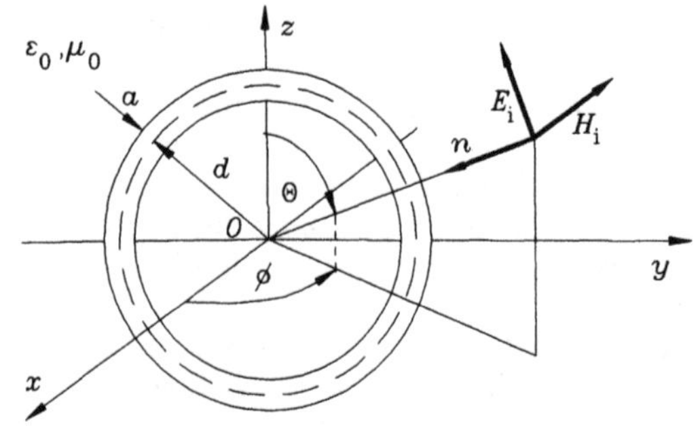

Figure 7.4 Sketch of a ring scatterer

Figure 7.5 shows the monostatic cross-section of the scatterer with $a/\lambda = 0.005$, plotted against the normalised radius d/λ of the ring. Two groups of the results are shown, $S_{mono}(\phi = 0°, \theta = 90°)/\lambda^2$ and $S_{mono}(\phi = 0°, \theta = 90°)/S_{mono}(\phi = 90°, \theta = 90°)$.

The results obtained by the present method are compared with the experimental results [93]. The results for S_{mono}/λ^2 obtained by subdomain approximation are not shown, because they coincide with those obtained by six-segment entire-domain approximation. However, the results obtained by the subdomain approximation cannot in any way be considered satisfactory in the second group of results, while those obtained by the entire-domain approximation are acceptable for both groups. This indicates that the entire-domain philosophy is also worthwhile with curved wires. If the ring is modelled

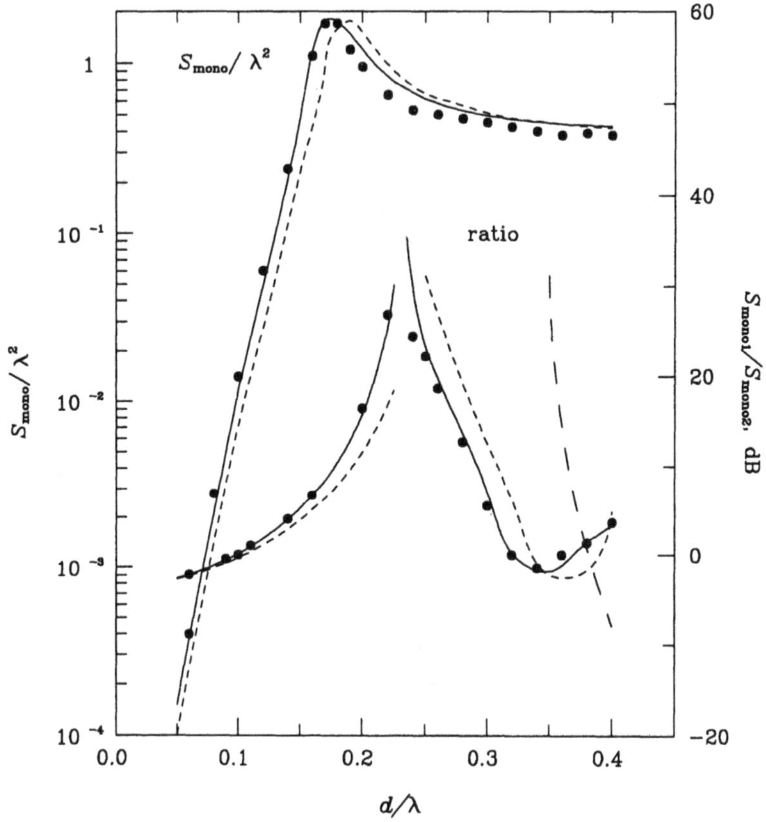

Figure 7.5 Monostatic cross-section of the ring scatterer sketched in Figure 7.4 against the ratio d/λ

$a/\lambda = 0.005$

———	generalised wire, entire domain
- - - - -	$S = 6$ segments, subdomain
. . . . . .	$S = 6$ segments, entire domain
● ● ●	experiment [93]

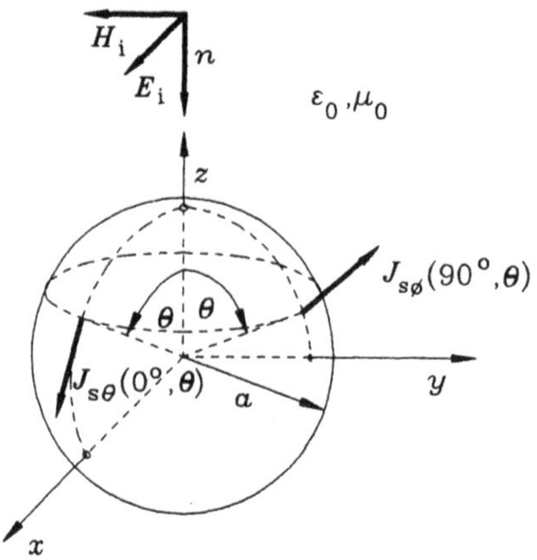

Figure 7.6 Sketch of a spherical scatterer

by cylinders, entire-domain approximation reduces the necessary number of subsegments when compared with the number indispensable with subdomain approximation. It is even more efficient to model the entire ring scatterer by a single generalised wire. In that case satisfactory results can be obtained for both groups of results with only six unknowns.

7.2.3 Approximation of a spherical scatterer by bilinear surface elements

Sketched in Figure 7.6 is a spherical scatterer. Figure 7.7 shows the moduli of the θ and ϕ components of the surface-current-density vector over the surface of the scatterer of radius $a = 0.2\lambda$, normalised with respect to the intensity of the incident magnetic field. The theoretical results were obtained by approximating the sphere with $M = 54$ bilinear surfaces, as in Figure 2.12*c*, and by using subdomain basis functions ($\mathcal{N} = 33$ unknowns). Indicated in Figure 7.7 is the current at the points in the middle of the sides of bilinear surfaces. These results are compared with theoretical results from Reference [49] ($\mathcal{N} = 144$ unknowns) and with the exact solution obtained by means of eigenfunctions. The results from Reference 49 were obtained for the sphere modelled by the same set of $M = 96$ triangles and for two orientations of this model of the sphere with respect to the direction of the incident wave. (The circles and triangles in Figure 7.7 indicate these two sets of results in both cases.) The results in Reference 49 are shown at the midpoints of the triangle sides. It is seen that the quality of the results obtained by the two approximate methods is very nearly the same.

7.2.4 Modelling of a spherical scatterer by generalised quadrilaterals

Figure 7.8 shows the same result as Figure 7.7, except that the sphere is modelled by six generalised quadrilaterals, as in Figure 2.5, and the current by entire-

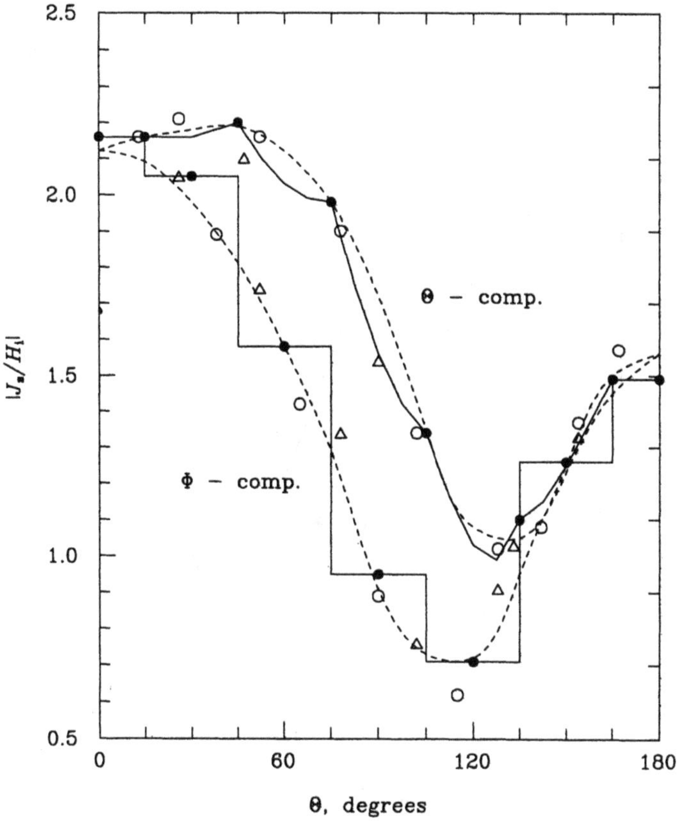

Figure 7.7 *Modulus of θ and φ components of the surface-current-density vector over the surface of the spherical scatterer sketched in Figure 7.6 normalised with respect to the intensity of the incident magnetic field*

Radius $a = 0.2\ \lambda$

—●— $M = 54$ bilinear surfaces, subdomain basis functions ($\mathcal{N} = 33$ unknowns), at the midpoints of the sides of bilinear surfaces (see Figure 2.12*c*)

- - - exact solution

○, △ theory ($\mathcal{N} = 144$ unknowns), at the midpoints of triangles [49]

domain approximation with a total of $\mathcal{N} = 42$ unknowns. These results are in excellent agreement with the exact results. It is therefore evident that for curved surfaces a procedure can be constructed which is significantly more accurate than that based on the approximation of a surface by bilinear (or triangular) surface elements. This was, of course, to be expected. Thus, for example, instead of bilinear surface elements bicubical surface elements can be used, i.e. curved surface elements defined by dual cubic splines, which can be defined, for example, by 4×4 points in space.

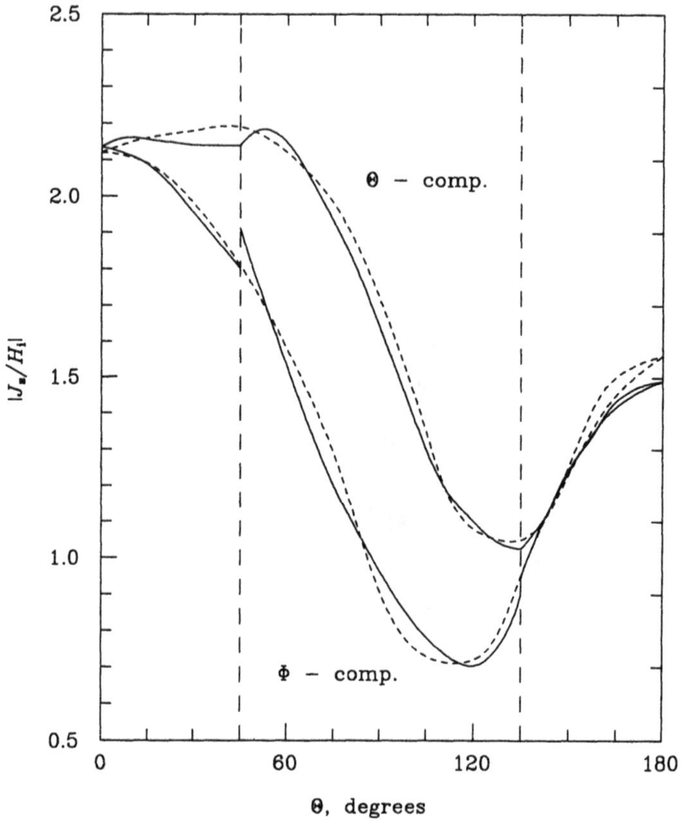

Figure 7.8 *The same result as in Figure 7.7, except that the sphere is modelled by six generalised quadrilaterals, as in Figure 2.5, and the current by entire-domain approximation with $N=42$ unknowns*

——— modelled by six generalised quadrilaterals
– – – exact results

7.2.5 *Monopole antenna at the centre of a square plate: attachment-mode approximation of wire-to-plate junction*

Figure 7.9 depicts a monopole antenna at the centre of a square plate. The monopole is driven by a TEM magnetic-current frill.

Figure 7.10 shows the conductance and susceptance of the monopole ($h = 421$ mm, $a = 0.8$ mm, $b = 4.7$ mm, $d = 350$ mm, $l = 914$ mm) plotted against frequency. The interconnection of the wire and the plate is modelled by two rings, one of radii a and b, and the second of radii b and d. Note that if the TEM magnetic-current frill were not present, a single ring, of radii a and d, would suffice. However, owing to the electric-field discontinuity at the frill edge, two rings are indispensable.

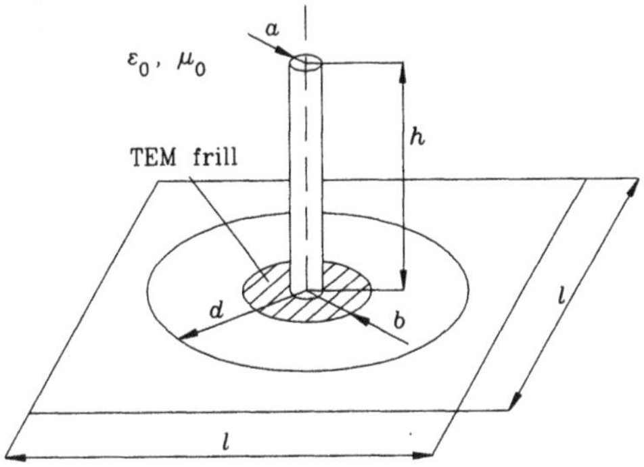

Figure 7.9 Monopole antenna at the centre of a square plate

The monopole is driven by a TEM magnetic-current frill. The interconnection of the wire and the plate is modelled by two rings, one of radii a and b, and the other of radii b and d

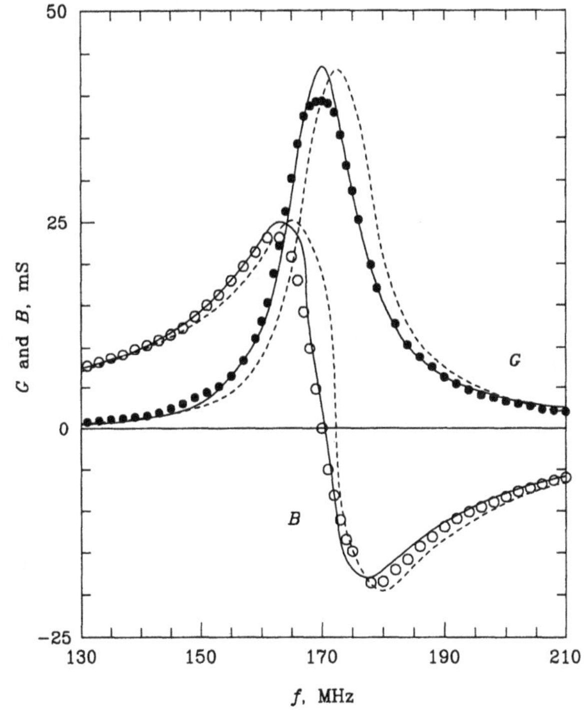

Figure 7.10 Conductance G and susceptance B of the monopole antenna sketched in Figure 7.9 against frequency

h=421 mm, a=0.8 mm, b=4.7 mm, d=350 mm, l=914 mm
—— this method, N=12 unknowns
- - - theory, N=43 unknowns [46]
○, ● experiment [46]

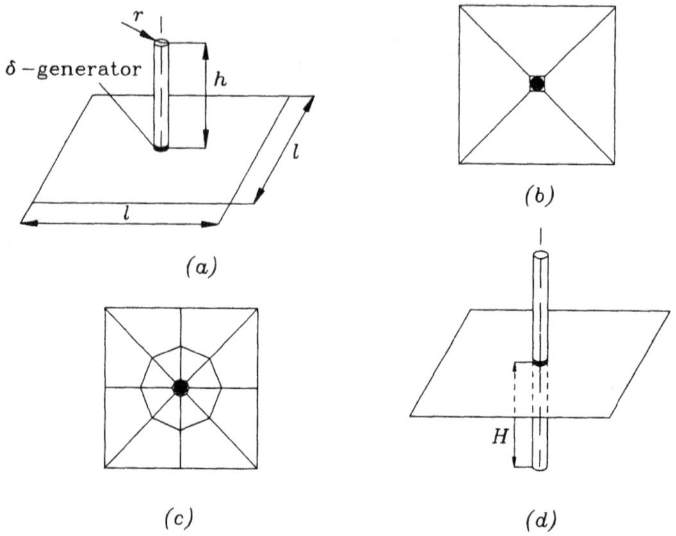

Figure 7.11 *Monopole antenna at the centre of a square plate*

a Driven by a delta-function generator at the monopole base
b Square plate divided into $M=4$ patches
c Square plate divided into $M=16$ patches
d Feeder of the monopole modelled by a wire segment on the opposite side of the plate

The theoretical results obtained by the proposed method, with $N=12$ unknowns, are compared with theoretical results (obtained with $N=43$ unknowns) and experimental results from Reference 46. The theoretical results from Reference 46 were obtained with a single attachment mode, as in Figure 2.14*a*. Good agreement is observed between the results obtained by the proposed method and the experimental results, while the other theoretical results appear to be significantly less accurate, in spite of a much larger number of unknowns.

7.2.6 Monopole antenna at the centre of a square plate: localised junction model

First, a monopole antenna is analysed at the centre of a square plate, as sketched in Figure 7.11*a*. It is assumed that the antenna is driven by a delta-function generator at the monopole base. The square plate is divided into (i) $M=4$ patches, and (ii) $M=16$ patches, as shown in Figures 7.11*b* and 7.11*c*.

Figure 7.12 shows the conductance and susceptance of the antenna ($h=421$ mm, $r=0.8$ mm and $l=914$ mm) plotted against frequency. The theoretical results obtained by the present method, with (i) $N=3$ unknowns ($M=4$ patches), and (ii) $N=21$ unknowns ($M=16$ patches), are compared with theoretical results obtained with $N=43$ unknowns and the

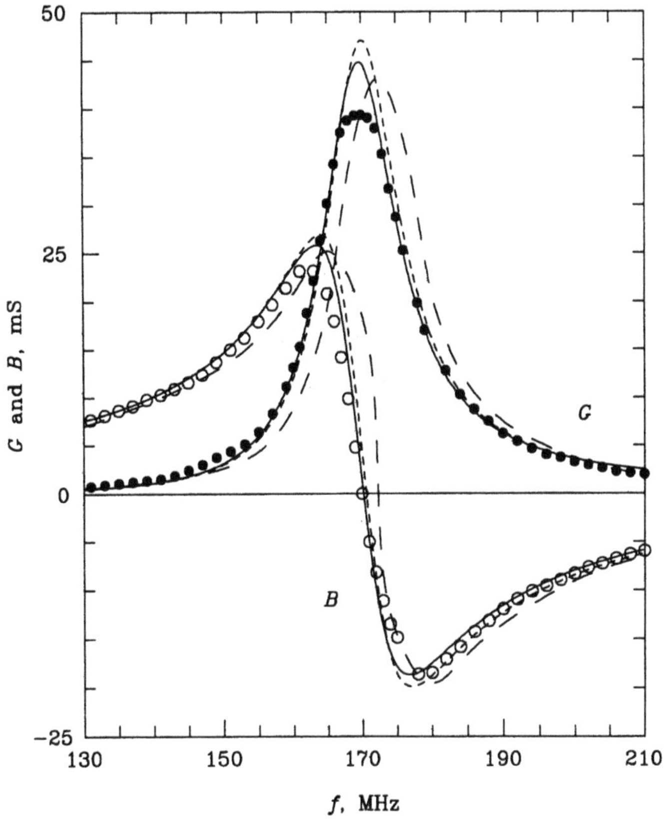

Figure 7.12 *Conductance G and susceptance B of the antenna sketched in Figure 7.11a against frequency*

$h = 421$ mm, $r = 0.8$ mm, $l = 914$ mm
—— this method, $N = 21$ unknowns ($M = 16$ patches)
······ this method, $N = 3$ unknowns ($M = 4$ patches)
— — theory, $N = 43$ unknowns [46]
●, ○, experiment [46]

experimental results from Reference 46. The theoretical results from Reference 46 were obtained by using the basic attachment mode shown in Figure 2.14*a*. Excellent agreement between the results obtained by the present method and the experimental results is observed, except in the vicinity of the resonant frequency. The theoretical results from Reference 46 are seen to be less accurate. The theoretical results obtained by the present method, with (i) $N = 11$ unknowns ($M = 4$ patches), and (ii) $N = 13$ unknowns ($M = 16$ patches) are practically identical to those obtained with $N = 21$ unknowns ($M = 16$ patches). If the number of patches and unknowns is further increased, the results stay practically unchanged. Hence,

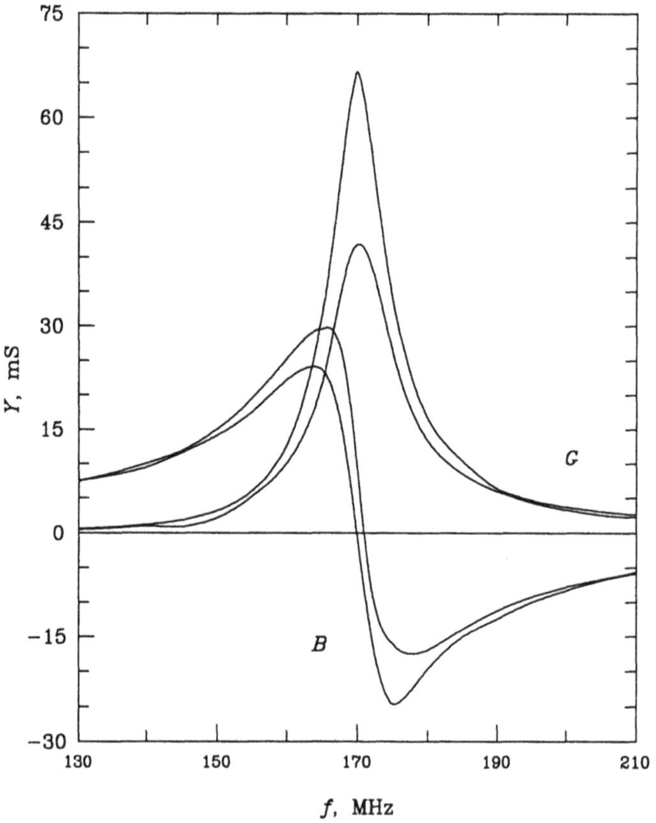

Figure 7.13 *Range of conductance and susceptance of the monopole sketched in Figure 7.11d with a wire approximating the feeder, against frequency, if the length H of the wire is varied between 0 and 1500 mm*

it can be concluded that the simple approximation of the plate sketched in Figure 7.11*b* can be considered satisfactory.

The difference between the theoretical and experimental results at the resonant frequency can probably be explained by the fact that the structures analysed are not identical to the experimental structure. For example, since the dimension of one side of the square plate is of the order of half a wavelength, the coaxial line used for the admittance measurements may influence the results of these measurements. In some cases this influence can be modelled by a wire on the opposite side of the plate, as sketched in Figure 7.11*d*. If the length H of this wire is changed from $H = 0$ to $H = 1500$ mm, the conductance and susceptance of the antenna are changed between minimum and maximum values as indicated in Figure 7.13. Very large differences between minimum and maximum values can be observed at the resonant frequency.

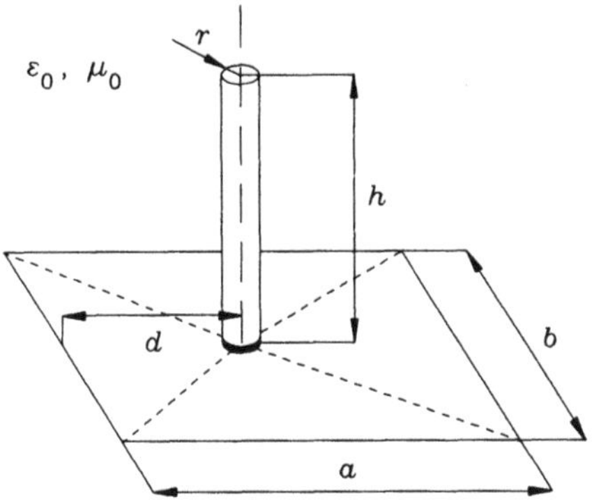

Figure 7.14 Monopole antenna mounted on a square plate

7.2.7 Monopole antenna near the edge of a square plate

Figure 7.14 shows a monopole antenna mounted on a square plate. It is assumed that the antenna is driven by a delta-function generator at the monpole base.

Figure 7.15 shows the resistance and reactance of the monopole with $h = \lambda/4$, $r = 0.001\lambda$ and $a = b = 1.2\,\lambda$ plotted against the normalised distance d/λ of the monopole axis from the edge of the plate. The theoretical results obtained by the present method, with $N = 54$ unknowns and $M = 4$ patches, are compared with theoretical results from Reference 58 obtained using a plate of different size ($a = 0.4\,\lambda$, $b = 0.5\,\lambda$), and with theoretical results from Reference 58 obtained for a halfplane. The first set of theoretical results from Reference 58 was obtained by using surface-patch modelling and edge-attachment mode indicated in Figure 2.14*b*, while the second set was obtained by using the exact Green functions for the monopole at the edge of a halfplane. Relatively good agreement is observed between the results obtained by the proposed method and the results obtained for a halfplane. The results obtained by using the edge-attachment mode are less accurate. This can be explained by the fact that the plate ($a = 0.4\lambda$, $b = 0.5\lambda$) is not large enough. By numerical experiments it was found that better agreement is obtained for plates with their minimum dimension greater than a wavelength.

7.2.8 Monopole antenna near the wedge formed by two square plates

Consider now a monopole antenna mounted near the wedge formed by two square plates, as sketched in Figure 7.16 It is assumed that the antenna is driven by a delta-function generator at the monopole base.

Figure 7.17 plots the monopole resistance and reactance against the normalised monopole distance d/λ from the wedge, for $h = 0.25\lambda$, $r = 0.0015\lambda$, $a = 0.4\lambda$ and $\alpha = 90°$. The theoretical results obtained by the

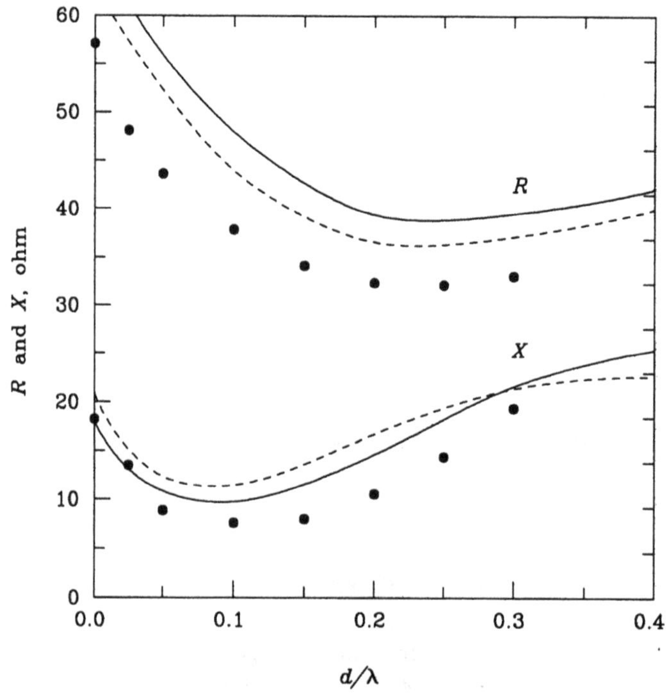

Figure 7.15 *Resistance R and reactance X of the monopole antenna sketched in Figure 7.14 against normalised distance d/λ of the monopole from the edge of the plate*

It is assumed that the antenna is driven by a delta-function generator at the monopole base
$h=\lambda/4$, $r=0.001\lambda$, $a=b=1.2\lambda$
——— this method, $N=54$ unknowns ($M=4$ patches)
- - - - exact, halfplane [58]
● ● ● theory, plate $a=0.4\ \lambda$, $b=0.5\ \lambda$ [58]

present method, with $N = 45$ unknowns and $M = 5$ patches, are compared with theoretical results obtained with $N = 29$ unknowns, and experimental results from Reference 59. The theoretical results from Reference 59 were obtained by using the wedge-attachment mode sketched in Figure 2.14*d*. Good agreement is found between the results obtained by the present method and experimental results, while the results for the reactance obtained with the wedge-attachment mode are significantly less accurate.

7.2.9 Monopole antenna near the vertex formed by three square plates

Figure 7.18 shows a monopole antenna near the vertex of three interconnected square metal plates. It is assumed that the antenna is driven by a delta-function generator at the monopole base.

Figure 7.19 shows the resistance and reactance of the monopole plotted against the normalised distance d/λ from two plate edges, for $h = 0.25\lambda$,

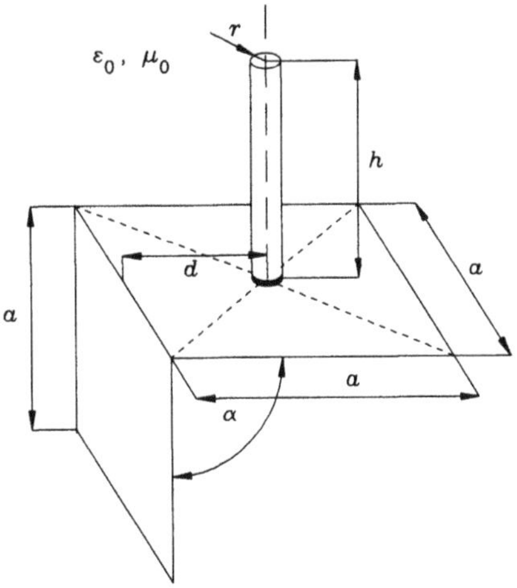

Figure 7.16 Monopole antenna mounted near a wedge formed by two square plates

It is assumed that the antenna is driven by a delta-function generator at the monopole base

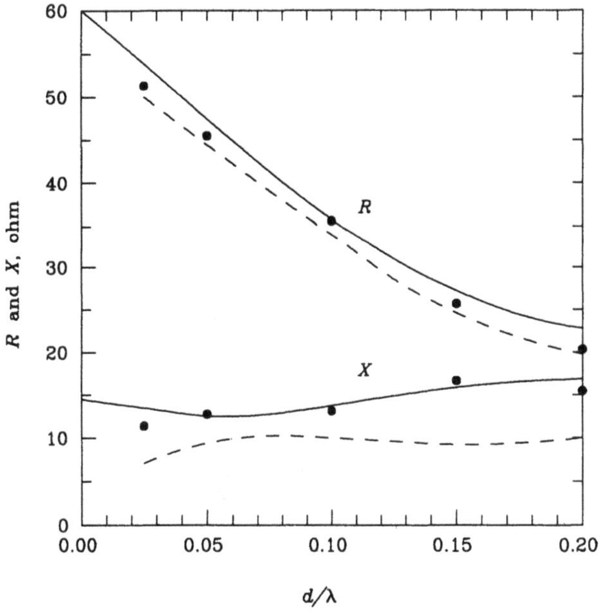

Figure 7.17 Resistance R and reactance X of the monopole antenna sketched in Figure 7.16 against normalised distance d/λ of the monopole from the wedge

$h=0.25\ \lambda$, $r=0.0015\ \lambda$, $a=0.4\ \lambda$, $\alpha=90°$

——— this method, $N=45$ unknowns ($M=5$ patches)

- - - - theory, $N=29$ unknowns [59]

● ● ● experiment [59]

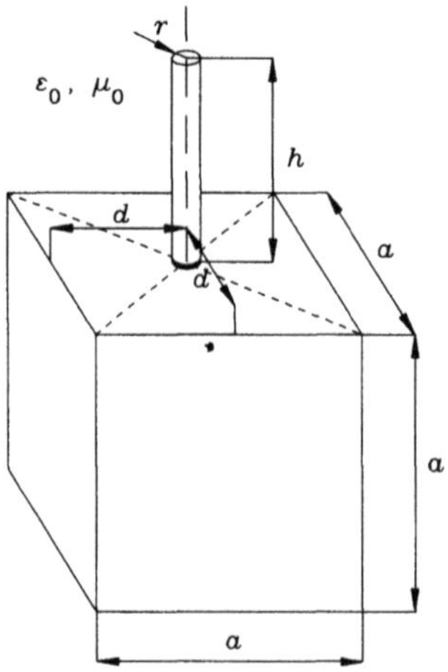

Figure 7.18 Monopole antenna near the vertex of three interconnected square metal plates

It is assumed that the antenna is driven by a delta-function generator at the monopole base

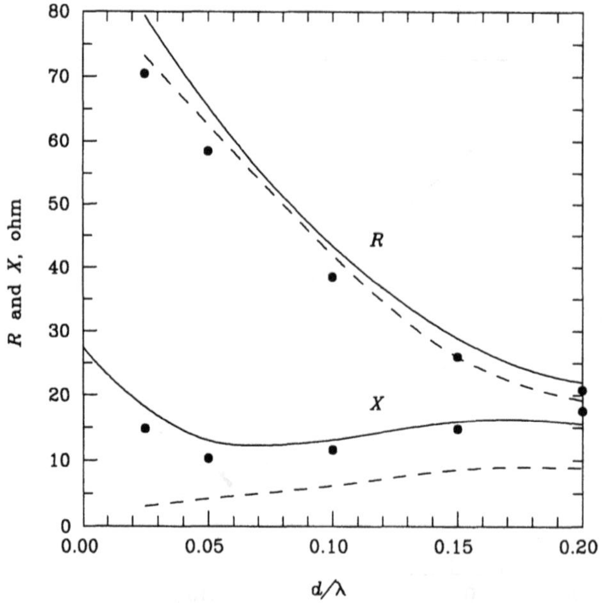

Figure 7.19 Resistance R and reactance X of the monopole antenna sketched in Figure 7.18 against the normalised distance d/λ of the monopole from the edges of the corner

$h=0.25\ \lambda$, $r=0.0015\ \lambda$, $a=0.4\ \lambda$

——— this method, $N=54$ unknowns ($M=6$ patches)

- - - - theory, $N=37$ unknowns [59]

● ● ● experiment [59]

$r = 0.0015\lambda$ and $a = 0.4\lambda$. The theoretical results obtained by the present method, with $N = 54$ unknowns and $M = 6$ patches, are compared with theoretical results obtained with $N = 37$ unknowns, and experimental results from Reference 58. The theoretical results from Reference 58 were obtained with the vertex-attachment mode (the simplified attachment mode sketched in Figure 2.14*c*). Very good agreement is evident between the results obtained by the proposed method and experimental results, while the results for the reactance obtained with the vertex-attachment mode are noticeably less accurate.

7.3 Examples illustrating the choice of the approximation of currents

In this Section, relative error in admittance of wire dipoles for various basis functions is considered (Sections 7.3.1 and 7.3.2), as well as accuracy in admittance of electrically long wire antennas obtained with polynomial and combined entire-domain approximations (Section 7.3.3). An electrically medium-sized rectangular scatterer is analysed as an example of surface structures (Section 7.3.4).

7.3.1 Resonant wire dipole: relative error in admittance for various basis functions

Consider the dipole antenna sketched in Figure 7.1. Assume that the dipole is a halfwave dipole. The aim of this Section is to analyse the error in the antenna admittance for various basis functions in the approximation of current along the dipole.

Figure 7.20 shows the relative error in the admittance of the dipole ($h/a = 100$, $b/a = 2.3$), plotted against the number of unknowns in the analysis. The relative error in the antenna admittance is defined as

$$\epsilon = \frac{|Y - Y_0|}{|Y_0|} \tag{7.1}$$

where Y_0 is the 'exact' value, obtained by several methods and with a sufficiently high degree of the current approximation. Convergence is improved by taking into account the effect of the flat dipole end. The results were obtained with the proposed entire-domain Galerkin method and with two subdomain Galerkin methods, the PWL (piecewise-linear) and the PWS (piecewise-sinusoidal). The advantage of the entire-domain analysis is evident, particularly if a more accurate analysis is required.

7.3.2 Antiresonant wire dipole: relative error in admittance for various basis functions

Consider now the same dipole as in Section 7.3.1 ($h/a = 100$, $b/a = 2.3$), except that its length is one wavelength. In Figure 7.21 the relative error in the admittance (defined as in eqn. 7.1) of the dipole is plotted against the number of unknowns in the analysis. The results were again obtained

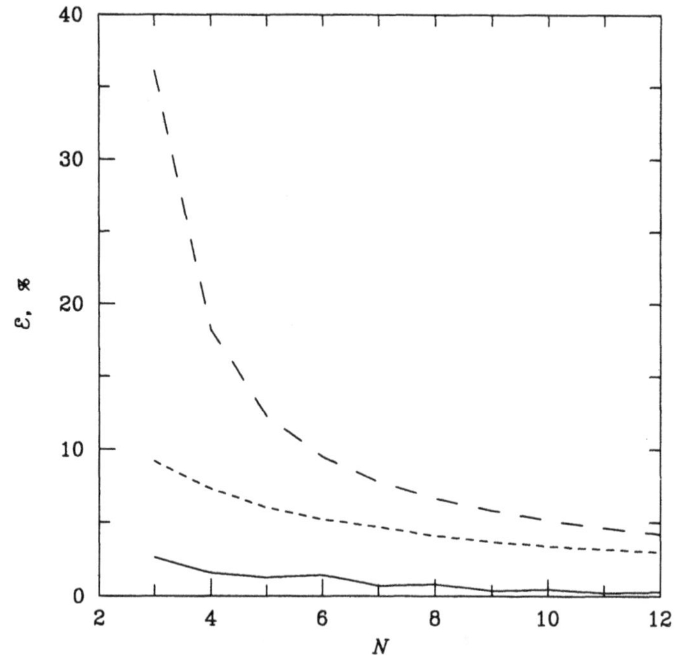

Figure 7.20 *Relative error in the admittance of a halfwave dipole against the number of unknowns in the analysis*

$h/a = 100$, $b/a = 2.3$

——— entire-domain Galerkin

— — — subdomain Galerkin (piecewise-linear)

- - - - - subdomain Galerkin (piecewise-sinusoidal)

by the entire-domain Galerkin method and two subdomain Galerkin methods, the PWL (piecewise-linear) and the PWS (piecewise-sinusoidal). The advantage of the entire-domain analysis is evident, particularly if a more accurate analysis is required.

7.3.3 Electrically long wire antenna

As the next example, consider an electrically long monopole wire antenna, the dipole counterpart of which is sketched in Figure 7.1, with $a = 0.021\lambda$ and $b/a = 2.3$. Figure 7.22 shows the conductance and susceptance of the monopole plotted against its electrical length. The results obtained with the entire-domain polynomial and with the entire-domain combined polynomial/trigonometric expansions, with $N = 10$ unknowns in both cases (one of which is for the end effect and one for the excitation zone), are compared with experimental results [81]. The results obtained with the polynomial expansions exhibit good agreement with experimental results up to about $\beta h = 15$, while those obtained with the combined polynomial/trigonometric expansion are accurate for the entire range of βh considered. (Of course, as stressed at the end of the

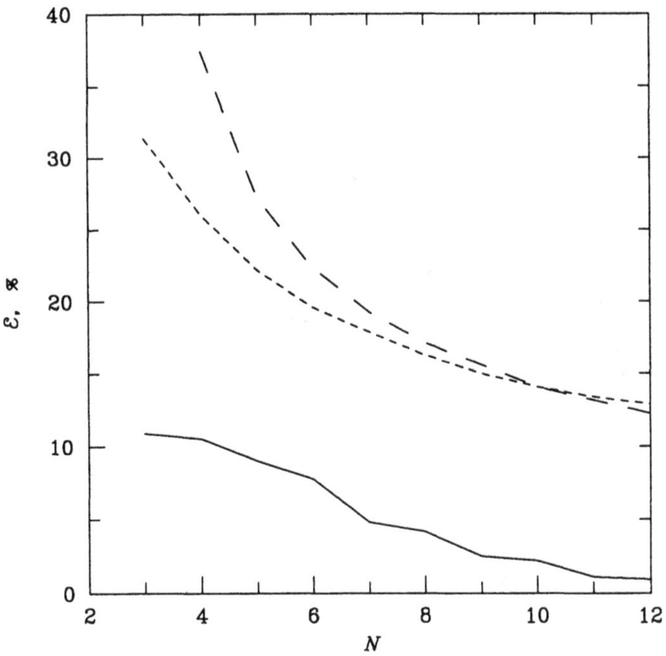

Figure 7.21 *Relative error in the admittance of a fullwave dipole against the number of unknowns in the analysis*

$h/a = 100$, $b/a = 2.3$
——— entire-domain Galerkin
– – – subdomain Galerkin (piecewise-linear)
- - - - - subdomain Galerkin (piecewise-sinusoidal)

introduction to this Chapter, the polynomial expansion can certainly also be used for longer antennas provided that the antennas are partitioned into smaller subsegments. What we desired, however, was a comparison of the expansions in the entire-domain approach.)

7.3.4 *Electrically medium rectangular scatterer*

Sketched in Figure 7.23 is an infinitely thin rectangular scatterer. Figure 7.24 shows the normalised radar cross-section of the scatterer, with $a = 2\lambda$ and $b = 3\lambda$ plotted against the angle ϕ. The normalisation coefficient is $S_{rad}(\phi = 90°, \theta = 90°) = 26.25\,$dB. The theoretical results obtained by the proposed method, with $N_1 = 34$ and $N_2 = 40$ unknowns for both current components, are compared with theoretical results obtained with $N = 75$ unknowns for the u component of the current alone and with experimental results from Reference 44. Reasonable agreement can be seen between theoretical and experimental results. These results, and some others, indicated

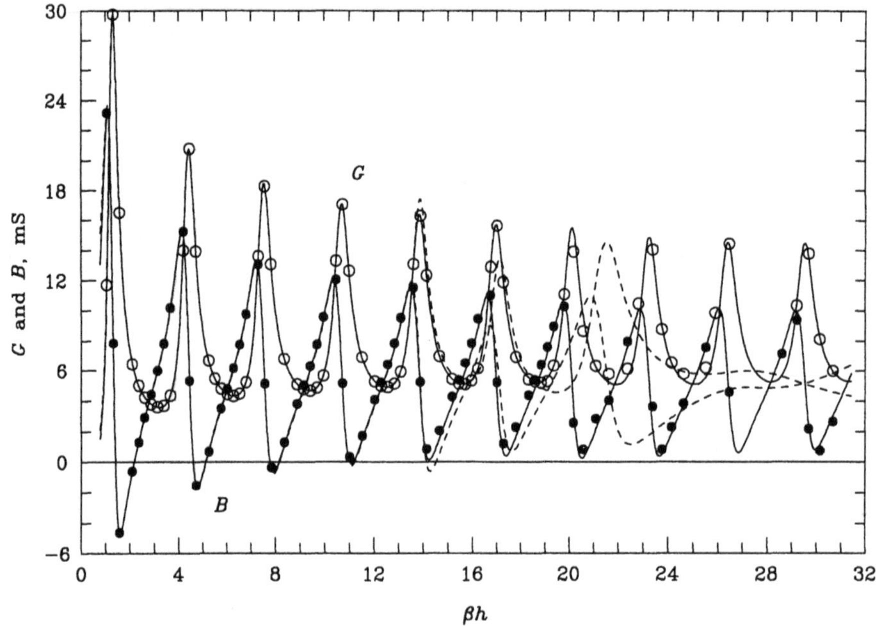

Figure 7.22 Conductance G and susceptance B of electrically long monopole wire antenna against the electrical length βh of its arm. See Figure 7.1 for its dipole equivalent

$a=0.021\ \lambda$, $b/a=2.3$
—— polynomial/trigonometric expansion, 10 unknowns
- - - polynomial expansion, 10 unknowns
○, ● experiment [81]

that, using the proposed method, it is sufficient for the approximation of both current components for large flat surfaces of simpler forms to use about ten unknowns per square of the wavelength.

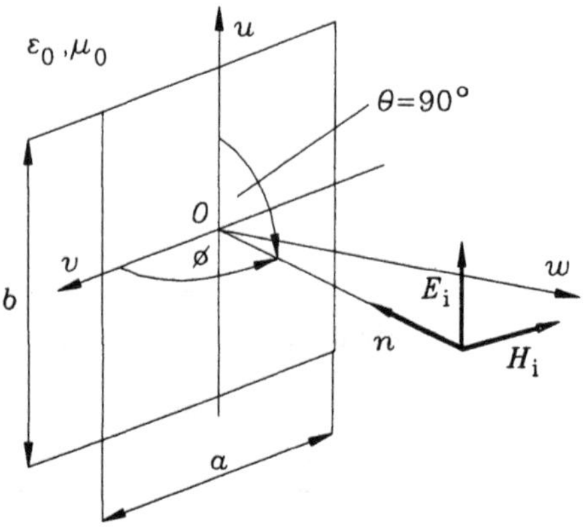

Figure 7.23 Sketch of an infinitely thin rectangular scatterer

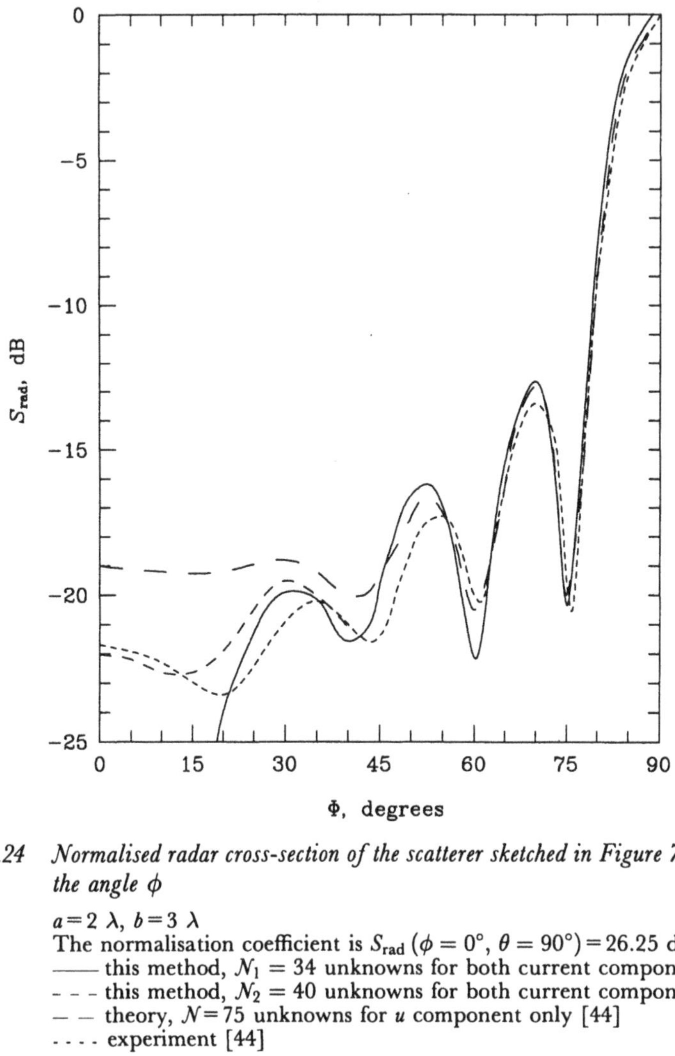

Figure 7.24 *Normalised radar cross-section of the scatterer sketched in Figure 7.23 against the angle ϕ*

$a=2\ \lambda$, $b=3\ \lambda$
The normalisation coefficient is S_{rad} ($\phi = 0°$, $\theta = 90°$) = 26.25 dB
—— this method, $\mathcal{N}_1$ = 34 unknowns for both current components
- - - this method, $\mathcal{N}_2$ = 40 unknowns for both current components
— — theory, $\mathcal{N}$=75 unknowns for u component only [44]
.... experiment [44]

7.4 Examples illustrating the modelling of excitation

This section presents two typical structures with various models of excitation. A resonant wire dipole is considered from that point of view in Section 7.4.1, and in Section 7.4.2 a monopole wire antenna above square plate.

7.4.1 Resonant wire dipole: various models of excitation

Figures 7.25 and 7.26 show the conductance and susceptance of the quarter-wavelength vertical monopole antenna ($h = \lambda/4$) above a perfectly conducting ground plane plotted against the order of polynomial approximation for the

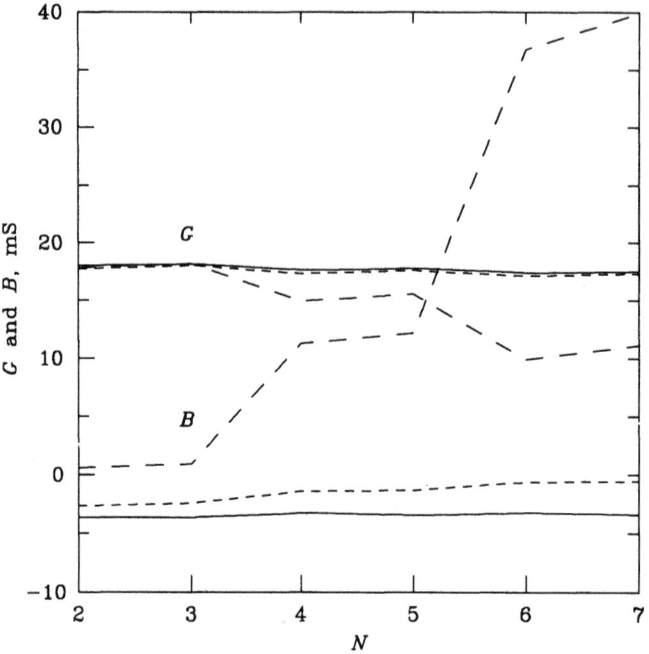

Figure 7.25 *Conductance G and susceptance B of quarter-wavelength vertical monopole antenna above perfectly conducting ground plane, against the order of polynomial approximation of current*

$h = \lambda/4$, $h/a = 10$
—— excitation by TEM magnetic-current frill, $b/a = 2.3$
– – excitation by delta-function generator
- - - - excitation by ring generator

current along the monopole. The monopole has the same symmetrical equivalent as that sketched in Figure 7.1, though for different excitations. The antenna is assumed to be excited by (i) the TEM magnetic-current frill (the ratio of outer-to-inner radius of the frill is $b/a = 2.3$), (ii) the delta-function generator and (iii) the ring generator. Figure 7.25 was obtained for the antenna height-to-radius ratio $h/a = 10$, and Figure 7.26 for $h/a = 30$. The results for $h/a = 100$ are not presented because they practically coincide for the three generator models, i.e. all generators give the same results for thin wires (h/a greater than 100) and a relatively low-order approximation for current (< 10). However, these figures suggest that for h/a less than about 100 more precise modelling of the excitation zone may be needed than is provided by a delta-function generator.

7.4.2 Monopole wire antenna above square plate: various models of excitation

As the next example, consider the monopole antenna sketched in Figure 7.27*a*. The antenna is analysed in four different ways:

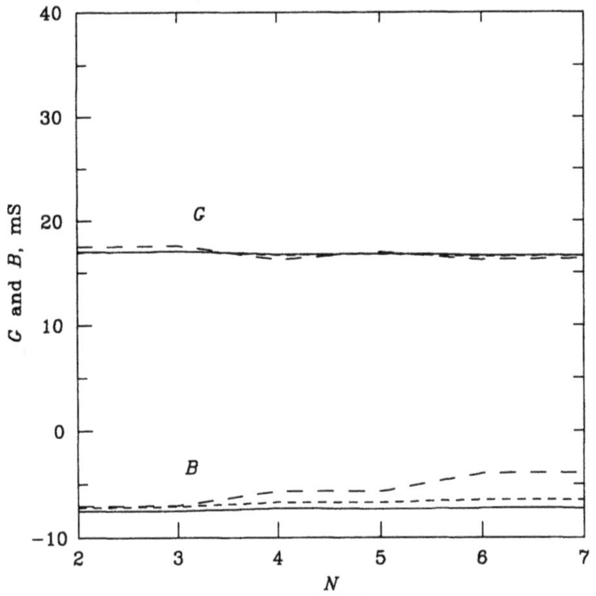

Figure 7.26 Conductance and susceptance of antenna as in Figure 7.25, for h/a = 30

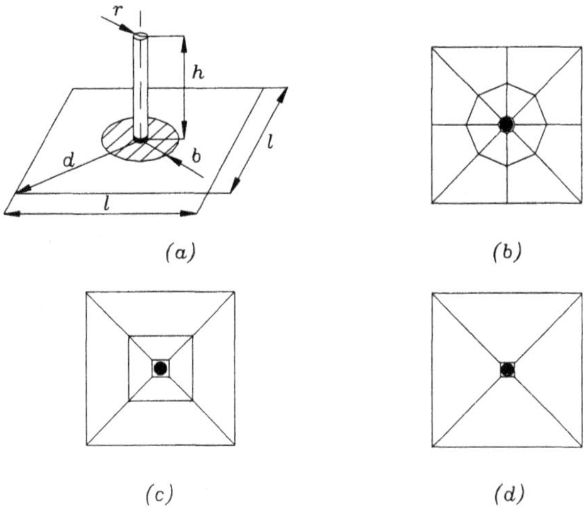

Figure 7.27 Monopole antenna above a square plate with different partitioning of the plate

a Antenna with original frill
b Plate divided into $M = 16$ patches, eight approximately coinciding with the frill
c Plate divided into $M = 8$ patches, four approximately coinciding with the frill
d Plate divided into $M = 4$ patches

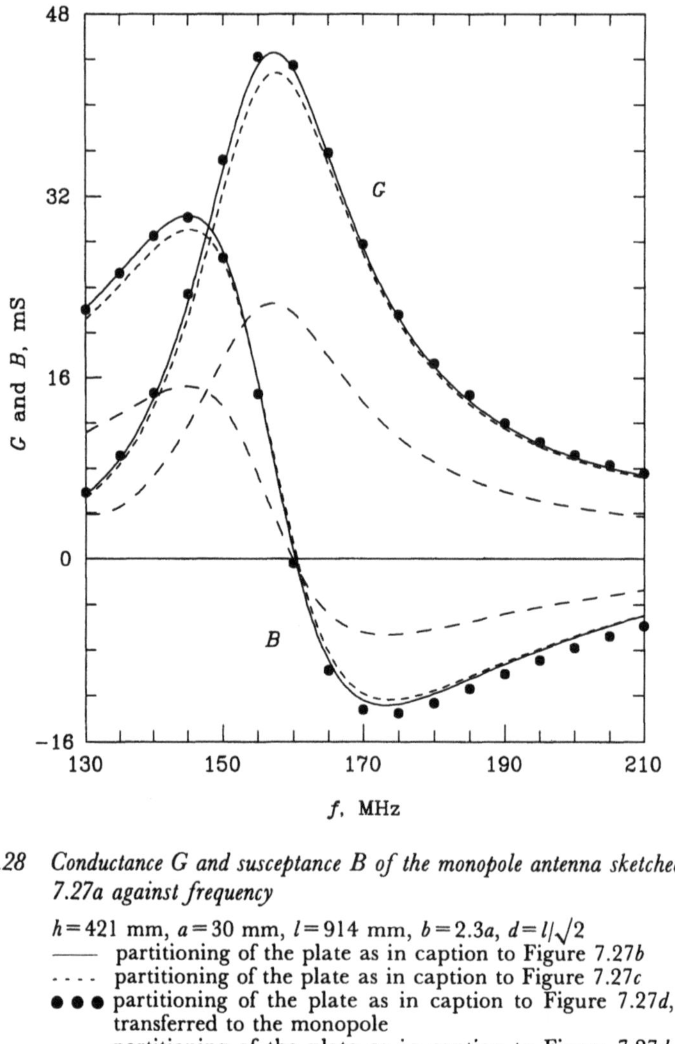

Figure 7.28 *Conductance G and susceptance B of the monopole antenna sketched in Figure 7.27a against frequency*

$h=421$ mm, $a=30$ mm, $l=914$ mm, $b=2.3a$, $d=l/\sqrt{2}$
—— partitioning of the plate as in caption to Figure 7.27*b*
.... partitioning of the plate as in caption to Figure 7.27*c*
●●● partitioning of the plate as in caption to Figure 7.27*d*, excitation transferred to the monopole
- - - partitioning of the plate as in caption to Figure 7.27*d*, excitation by frill with no attachment modes

(i) The plate is divided into $M = 16$ patches, eight of which approximately coincide with the frill, as indicated in Figure 7.27*b*.
(ii) The plate is divided into $M = 8$ patches, four of which approximately coincide with the frill, as indicated in Figure 7.27*c*.
(iii) The plate is divided into $M = 4$ patches, as indicated in Figure 7.27*d*, and the frill is replaced by the equivalent impressed electric field along the monopole.
(iv) The plate is divided into $M = 4$ patches as in Figure 7.27*d*, which overlap the frill, but neither further plate partitioning nor an attachment mode is used for more precise modelling of the frill.

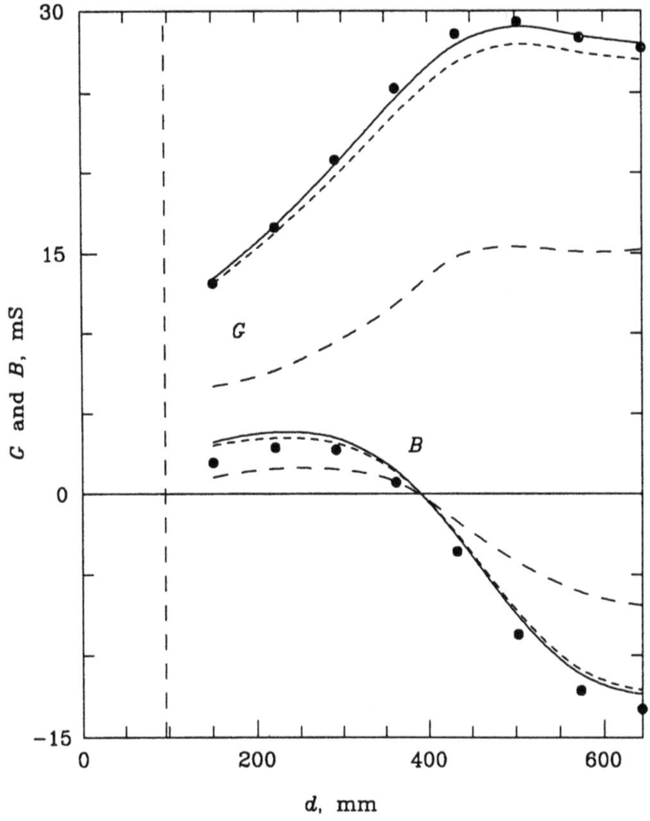

Figure 7.29 *Conductance G and susceptance B of the monopole antenna sketched in Figure 7.27a at f = 170 MHz against the distance d*

$h=421$ mm, $a=30$ mm, $l=914$ mm, $b=2.3a$
—— partitioning of the plate as in caption to Fig. 7.27*b*
.... partitioning of the plate as in caption to Fig. 7.27*c*
● ● ● partitioning of the plate as in caption to Fig. 7.27*d*, excitation transferred to the monopole
- - - partitioning of the plate as in caption to Fig. 7.27*d*, excitation by frill with no attachment modes

Figure 7.28 shows the conductance and susceptance of the antenna ($h = 421$ mm, $a = 30$ mm, $l = 914$ mm, $b = 2.3a$, $d = l/\sqrt{2}$), plotted against frequency. It is seen that in the first three cases the results practically coincide, while in the last case deviation is quite large, as should be expected. This shows that frill excitation requires additional segmentation, while excitation by an equivalent impressed field along the monopole does not.

Shown in Figure 7.29 is the conductance and susceptance of the monopole antenna sketched in Figure 7.27*a* ($h = 421$ mm, $a = 30$ mm, $l = 914$ mm, $b = 2.3a$), at a frequency $f = 170$ MHz, plotted against the distance d. The

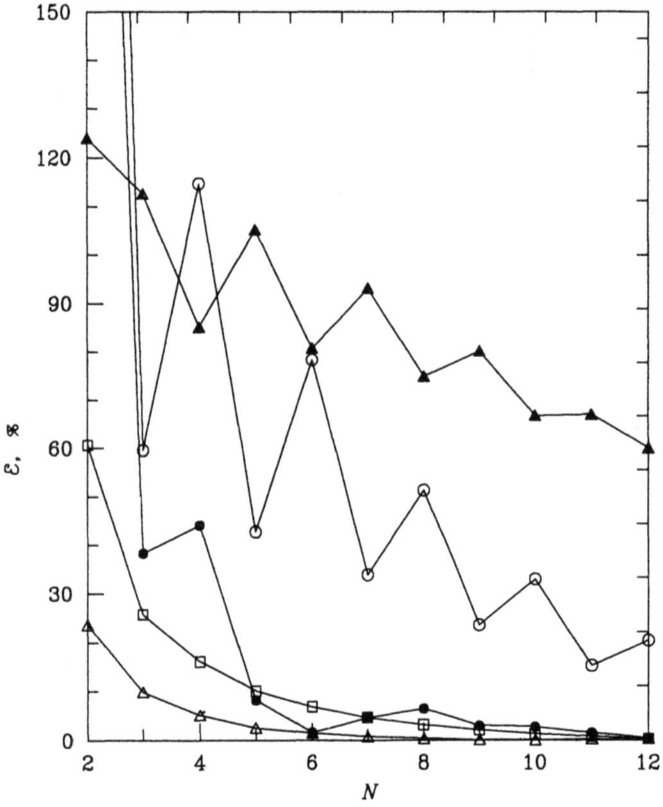

Figure 7.30 *Relative error in admittance of the halfwave dipole sketched in Figure 7.1 against the number of unknowns in the analysis*

$h/a = 100$, $b/a = 2.3$

○ ○ ○ point-matching, equidistant matching points

● ● ● point-matching, matching points at zeros of Gauss–Legendre polynomials

△ △ △ Galerkin

▲ ▲ ▲ least-squares

□ □ □ testing by subdomain sine functions

vertical broken line in the Figure indicates the distance *d* for which the frill reaches the plate edges. The antenna is analysed with all four models of excitation. It is seen that the equivalent impressed field along the monopole gives very good results even when the antenna is situated very close to the plate edge.

7.5 Examples illustrating the choice of test procedure

To illustrate the reasons for adopting a specific test procedure (the Galerkin method) in this book a number of examples aimed at justifying this choice are given in this Section. Thus using different test procedures, a resonant wire dipole is analysed in Section 7.5.1, an antiresonant wire dipole is analysed in Section 7.5.2, a resonant wire dipole is analysed for different dipole thicknesses in Section 7.5.3, and an analysis of a surface structure (a hexagonal plate scatterer) is presented in Section 7.5.4 from this point of view.

7.5.1 Resonant wire dipole: analysis by different test procedures

Figure 7.30 shows the relative error in admittance of a halfwave dipole sketched in Figure 7.1 ($h/a = 100$, $b/a = 2.3$), plotted against the number of unknowns in the analysis. The relative error is calculated according to eqn. 7.1. The results were obtained using the following five test procedures: the point-matching method with equidistant matching points; the point-matching method with matching points at the zeros of the Gauss–Legendre polynomials; the least-squares method; the Galerkin method; and testing by subdomain sine functions (which corresponds to solving the Hallén equation by means of point-matching [94]). To make the comparison between the point-matching method and the other test procedures more realistic, the quasistatic condition at the excitation point was added to the optimal set of conditions for the currents. (Note that this increased the error in the last two test procedures.) It is seen that the Galerkin method is somewhat better than testing by subdomain sine functions, and considerably better than the other test procedures.

7.5.2 Antiresonant wire dipole: analysis by different test procedures

Figure 7.31 shows the relative error in the admittance of a fullwave dipole ($h/a = 100$, $b/a = 2.3$) plotted against the number of unknowns in the analysis. Testing was performed by the same five methods as in the preceding example. It is seen that the best results are obtained with the point-matching procedure with matching points at the zeros of the Gauss–Legendre polynomials, the Galerkin method and the least-squares method. The error is on average significantly larger than for the resonant dipole of Section 7.5.1. This was to be expected, because the error for the resonant dipole is evaluated at the point where the current intensity is approximately a maximum, and for the antiresonant dipole is evaluated at the point where it is approximately a minimum. Note that both the results obtained by subdomain testing and those obtained by the Galerkin method would have been significantly better if the quasistatic condition at the excitation point had been omitted.

7.5.3 Resonant wire dipole: analysis by different test procedures and for different dipole thicknesses

Shown in Figure 7.32 is the relative error in the admittance (defined as in eqn. 7.1) of a resonant dipole antenna, of the form shown in Figure 7.1 and with $b/a = 2.3$, plotted against h/a. The analysis was performed for the same test procedures as in Section 7.5.1, by means of the EFIE and polynomial current expansion of degree $n = 8$. It is seen that the Galerkin method is not sensitive to the ratio h/a, that testing with subdomain functions is weakly sensitive to h/a, while the other test procedures, particularly the least-squares procedure, are very

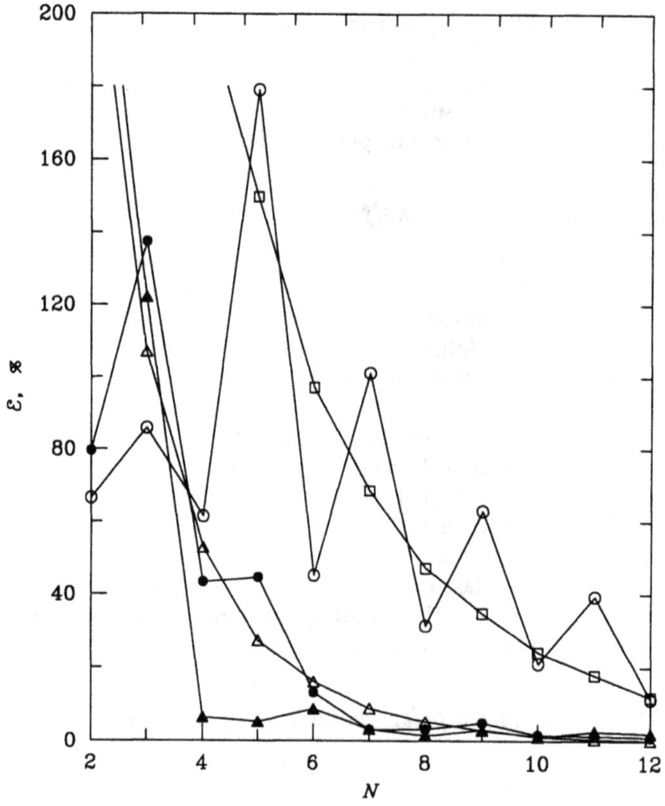

Figure 7.31 *Relative error in the admittance of a fullwave dipole against the number of unknowns in the analysis*

$h/a = 100$, $b/a = 2.3$

○○○ point-matching, equidistant matching points

●●● point-matching, matching points at zeros of Gauss–Legendre polynomials

△△△ Galerkin

▲▲▲ least-squares

□□□ testing by subdomain sine functions

sensitive to changes in h/a. This means that, for larger ratios h/a, it is necessary to increase the number of unknowns considerably.

7.5.4 Hexagonal-plate scatterer: analysis by different test procedures

As the last example, consider an electrically small equilateral hexagonal scatterer of side a in a vacuum as shown in Figure 7.33. The scatterer is excited by a uniform plane electromagnetic wave normally incident on the scatterer surface. The incident wave is polarised in the direction of the u axis.

The scatterer was modelled by two trapezoids, obtained by dividing the scatterer along the v axis. The analysis was performed by solving the EFIE by

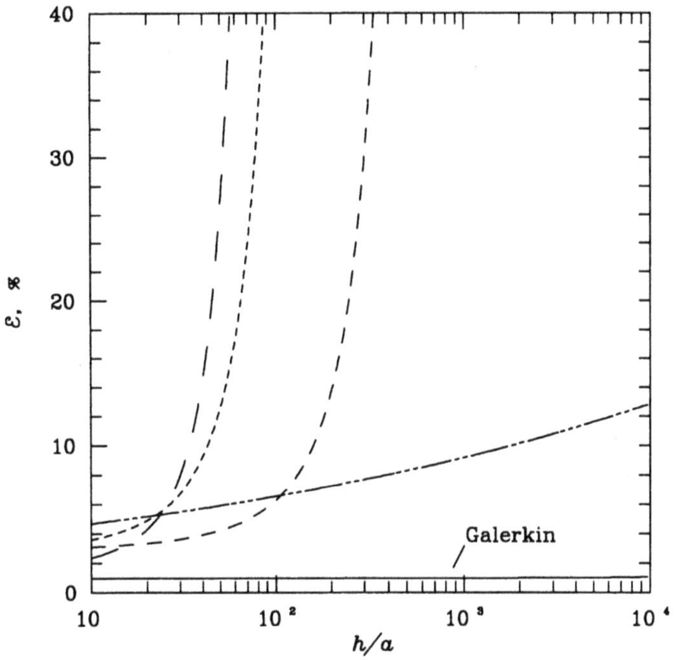

Figure 7.32 Relative error in admittance of a halfwave dipole against h/a for different test procedures

The analysis was performed with polynomial of degree $N=8$ and the EFIE $b/a=2.3$

- - - - point-matching, equidistant matching points
---- point-matching, matching points at arguments of the Gauss–Legendre integration formula
—— Galerkin
— — least-squares
- - — testing by subdomain sine functions

the Galerkin method and with different numbers of matching points (in the extended sense): $6 \times 3, 8 \times 4$ and 12×6. (Note that the first of these corresponds approximately to the point-matching method.) Each component of the surface

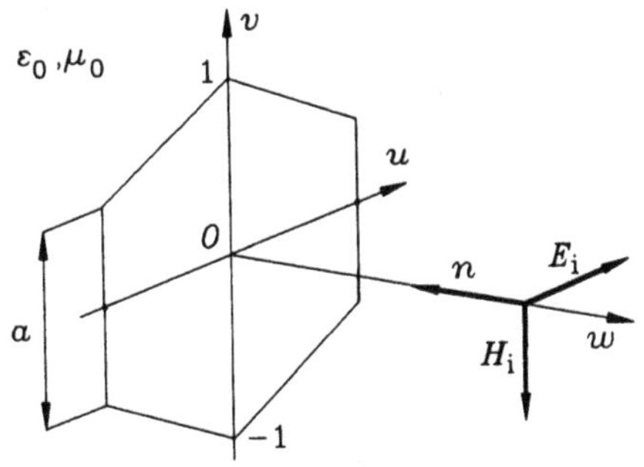

Figure 7.33 Sketch of an electrically small equilateral hexagonal scatterer

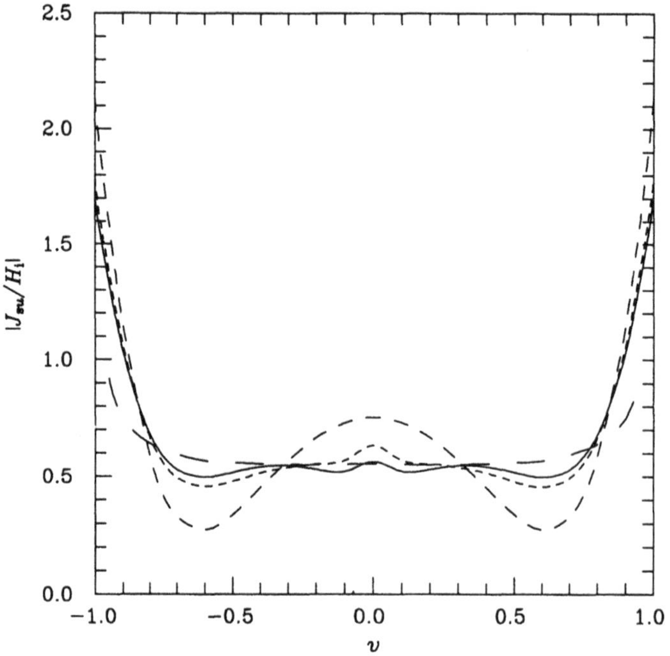

Figure 7.34 *Normalised magnitude $|J_{su}(v, 0)/H_i|$ of the u component of the surface-current-density vector along the v axis for the hexagonal scatterer sketched in Figure 7.33; $a = 0.0536\ \lambda$*

– – quasistatic solution for a disc of $a = 0.05\ \lambda$
- - - Galerkin, 6×3 matching points
. . . . Galerkin, 8×4 matching points
—— Galerkin, 12×6 matching points

current over the two trapezoids was approximated by a double polynomial of degree four, i.e. it was chosen that $n_u = n_v = 4$. As symmetry was taken into account, the total number of unknowns was $N = 16$.

Figure 7.34 shows the magnitude of the u component of the surface-current-density vector along the v axis, normalised with respect to the intensity of the incident magnetic field, i.e. the ratio $|J_{su}(v, 0)/H_i|$, for a scatterer with $a = 0.0536\lambda$. These results are compared with the quasistatic solution for a disc of radius $a = 0.05\lambda$. It is seen that for a stable solution the number of matching points needs to be several times greater than the theoretical minimum number of matching points. Similar conclusions were arrived at by increasing the degree of approximation in this example, or in the analysis of some other examples. Consequently, it seems that the point-matching procedure cannot be used to obtain stable results in the general case. It can also be concluded that, of the three test procedures considered (point-matching, Galerkin and least-squares), the Galerkin method appears to be the optimum.

Chapter 8

Numerical examples illustrating possibilities of the method

8.1 Introduction

In this Chapter various examples are presented intended to illustrate some possibilities of the method proposed in the book for the analysis of metallic antennas and scatterers. The method being quite general, it was possible to choose only a limited number of examples, but it is hoped that they will indicate the advantages and possible shortcomings of the proposed method when compared with other methods.

If not stated otherwise, the results presented in this Chapter were obtained by the polynomial approximation (satisfying the continuity equation at interconnections and the quasistatic condition at conical ends, which we shall term conditionally 'the optimum set of conditions') and the Galerkin method. Symmetry was used whenever possible to reduce the number of unknowns. Also, if not stated otherwise, the general localised-junction model was used for the treatment of wire-to-plate junctions.

8.2 Wirelike antennas

This Section presents the results for antennas which fall basically in the thin-wire antenna category. A monopole antenna with a compensating parasitic element is

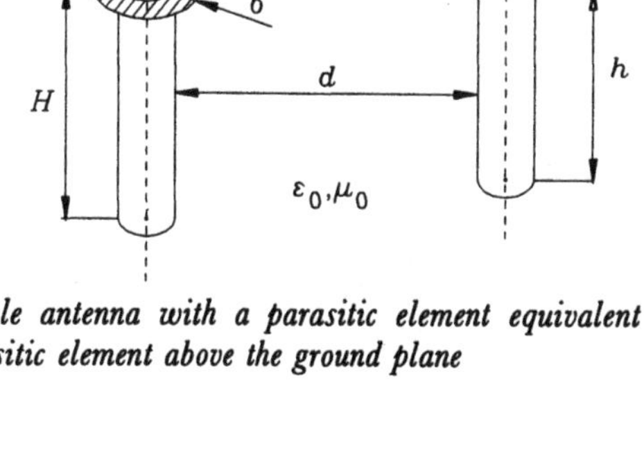

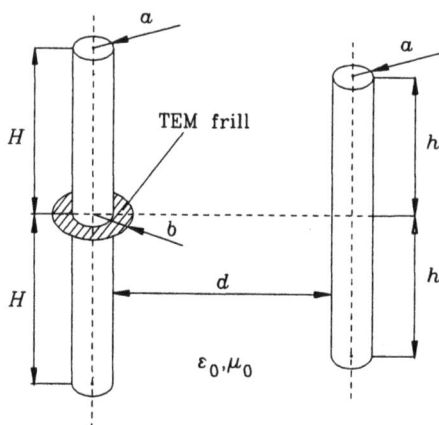

Figure 8.1 Dipole antenna with a parasitic element equivalent to a monopole with a parasitic element above the ground plane

analysed in Section 8.2.1. In Section 8.2.2 a monopole antenna with a compensating branch is analysed. Section 8.2.3 treats a small Yagi–Uda array, Section 8.2.4 a conical antenna, and Section 8.2.5 a monopole wire antenna above a circular metal plate.

8.2.1 Monopole antenna with parasitic element

Consider a vertical thin-wire monopole antenna with a parasitic element, driven at the ground plane by a coaxial line. The image theory yields an (approximately) equivalent system in the form of a dipole centre-driven by a TEM magnetic-current frill, with a parallel parasitic element, as shown in Figure 8.1. The antenna was analysed in two ways: not taking the end effect into

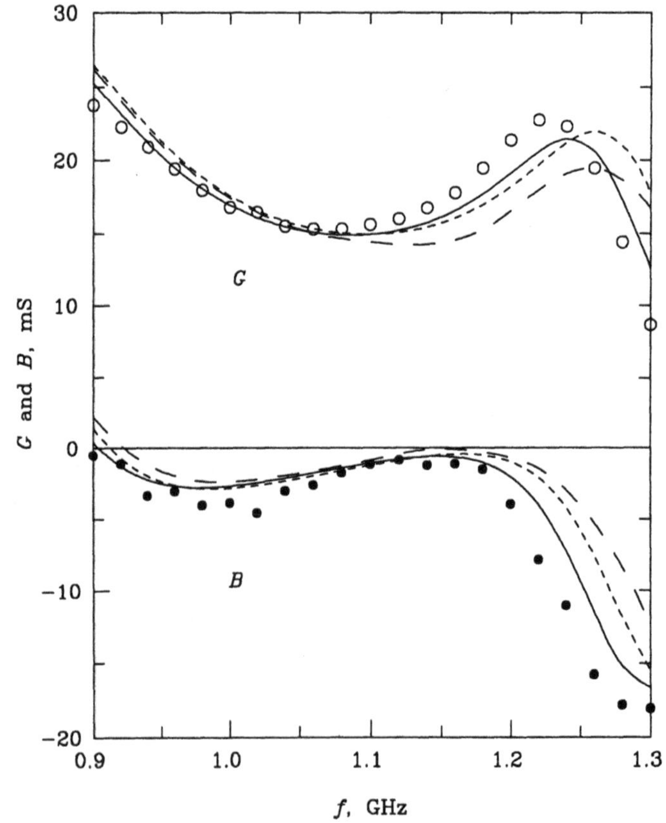

Figure 8.2 Conductance G and susceptance B of the monopole antenna the symmetrical equivalent of which is sketched in Fig. 8.1 against frequency

$H=79.2$ mm, $h=50$ mm, $d=18.5$ mm, $a=3$ mm, $b=6.9$ mm
—— this method, seven unknowns, with end effect
. . . . this method, four unknowns, without end effect
– – – theory, 13 unknowns [95]
○, ● experiment [95]

account (requiring four unknowns), and with taking the end effect into account (requiring seven unknowns).

Figure 8.2 shows the conductance and susceptance of the antenna with $H = 79.2$ mm, $h = 50$ mm, $d = 18.5$ mm, $a = 3$ mm and $b = 6.9$ mm, plotted against frequency. Also shown are theoretical and experimental results for the same antenna from Reference 95, where theoretical results were obtained with $N = 13$ unknowns. Note that the theoretical results from Reference 95 do not properly take into account the end effect. Very good agreement is observed between the experimental results and the results obtained by the method adopted in this book (with the end effect taken into account).

8.2.2 *Wire antenna with a compensating branch*

Figure 8.3 shows a monopole antenna with a compensating branch, driven by a coaxial line. It is adopted that $h = 81.5$ mm, $l = 54.2$ mm, $d = 9.5$ mm, $a = 3$ mm and $b/a = 2.3$.

Figure 8.4 shows the conductance and susceptance of the antenna, plotted against frequency. The antenna was analysed without taking the end effect into account (requiring five unknowns), and with the end effect taken into account (requiring eight unknowns). The results obtained using the proposed method are compared with theoretical results ($N = 13$ unknowns) and experimental results from Reference 33. (Note that the theoretical results from Reference 33 do not properly take into account the end effect.) As expected, if the end effect is taken into account much more accurate results are obtained.

8.2.3 *Yagi–Uda antenna*

Sketched in Figure 8.5 is a small Yagi–Uda array ($h_1 = 73.3$ mm, $h_2 = h_3 = 66.5$ mm, $h_4 = 64.3$ mm, $h_5 = 79$ mm, $d_1 = 43$ mm, $d_2 = 86$ mm, $d_3 = 129$ mm, $d_4 = 63$ mm, $a = 3$ mm, $b/a = 2.3$), driven by a coaxial line.

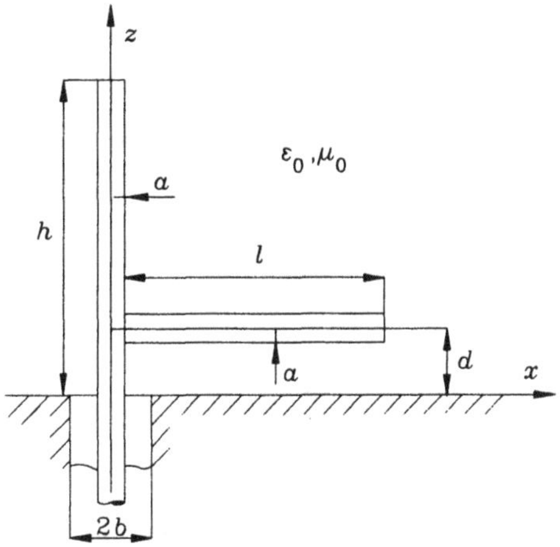

Figure 8.3 Monopole antenna with a compensating branch

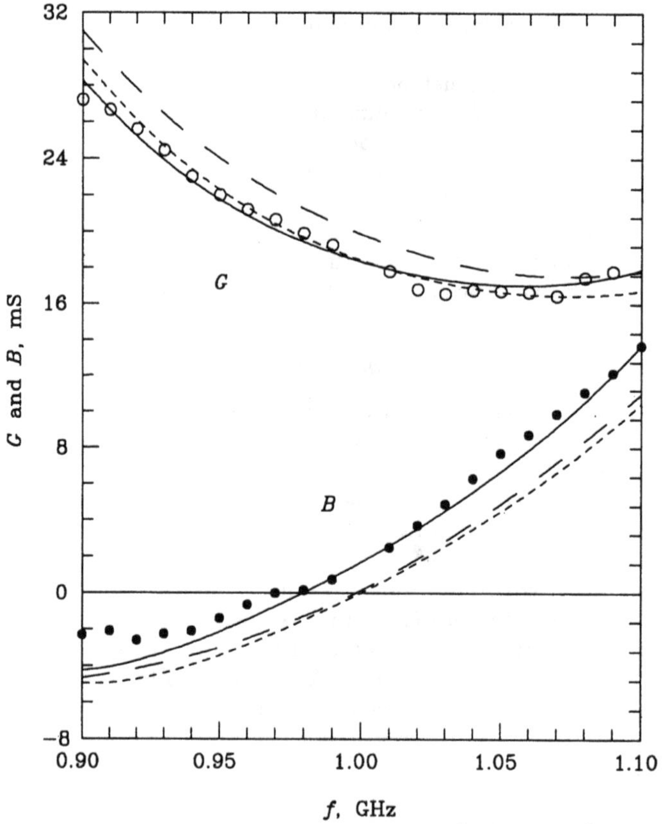

Figure 8.4 *Conductance G and susceptance B of the monopole antenna with compensating branch sketched in Figure 8.3 against frequency*

$h = 81.5$ mm, $l = 54.2$ mm, $d = 9.5$ mm, $a = 3$ mm, $b/a = 2.3$
—— this theory, eight unknowns, with end effect
- - - - this theory, five unknowns, without end effect
– – – theory, 13 unknowns [33]
○, ● experiment [33]

Figure 8.6 shows the antenna conductance and susceptance plotted against frequency. The antenna was analysed without the end effect taken into account (requiring 10 unknowns) and with (requiring 15 unknowns). The results obtained using the proposed method are compared with the theoretical results ($N = 26$ unknowns) and experimental results from Reference 33. (Note that the theoretical results from Reference 33 do not properly take into account the end effect.) It is seen that when the end effect is taken into account remarkably better agreement with experimental results is obtained.

8.2.4 Conical antenna

Figure 8.7 shows a conical monopole antenna above a ground plane ($h = 1.5\,\text{mm}$, $H = 88.6\,\text{mm}$, $D = 80\,\text{mm}$, $\alpha = 30°$, $a = 3\,\text{mm}$, $b = 6.9\,\text{mm}$).

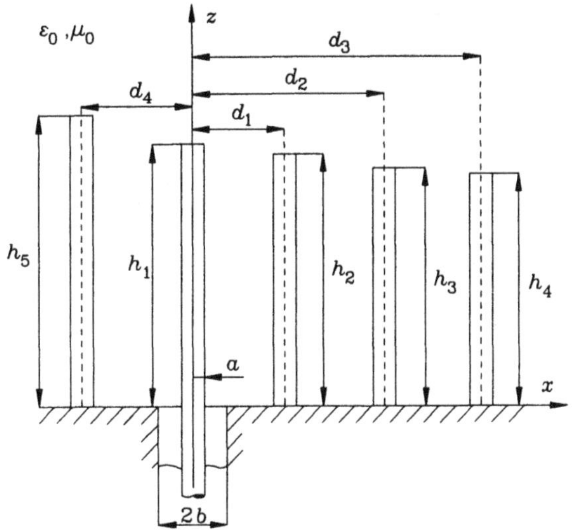

Figure 8.5 Sketch of a small Yagi–Uda array

The antenna was analysed as a symmetrical antenna centre-driven by a TEM magnetic-current frill. The dependence of the antenna conductance and susceptance on frequency is sketched in Figure 8.8. Good agreement can be observed between the results obtained by the proposed method ($\mathcal{N} = 6$ unknowns) and the experimental results from Reference 77. The theoretical results from Reference 77 ($\mathcal{N} = 15$ unknowns) are not presented, because they almost precisely coincide with the results obtained by the proposed method.

8.2.5 Monopole wire antenna above circular plate

Sketched in Figure 8.9 is a thin-wire monopole antenna fed at the centre of a circular plate by a coaxial line ($h = h_0$, h_1 or h_2, $h_0 = 0.224\lambda$, $h_1 = 0.229\lambda$, $h_2 = 0.2255\lambda$, $a = 0.003\lambda$, $b/a = 2.3$). In the first approximation the presence of the coaxial line can be neglected, and the antenna assumed to be driven by a TEM magnetic-current frill. The circular plate was modelled as two flat rings, one of radii a and b, and the other of radii b and d. Shown in Figure 8.10 is the antenna conductance and susceptance plotted against the normalised diameter of the plate with respect to the wavelength. The total number of unknowns amounted to $\mathcal{N} = 6$.

The results obtained by the present method ($\mathcal{N} = 6$ unknowns) are compared with the theoretical results ($\mathcal{N} = 14$ unknowns) and the experimental results from Reference 96. (Note that in Reference 96 the end effect was taken into account by adding approximately the equivalent length to the monopole height determined by numerical experiments. This is the meaning of the three slightly different monopole heights h_0, h_1 and h_2.) The conductance results for $h = h_2$ obtained by the present method are not shown, being almost identical to those

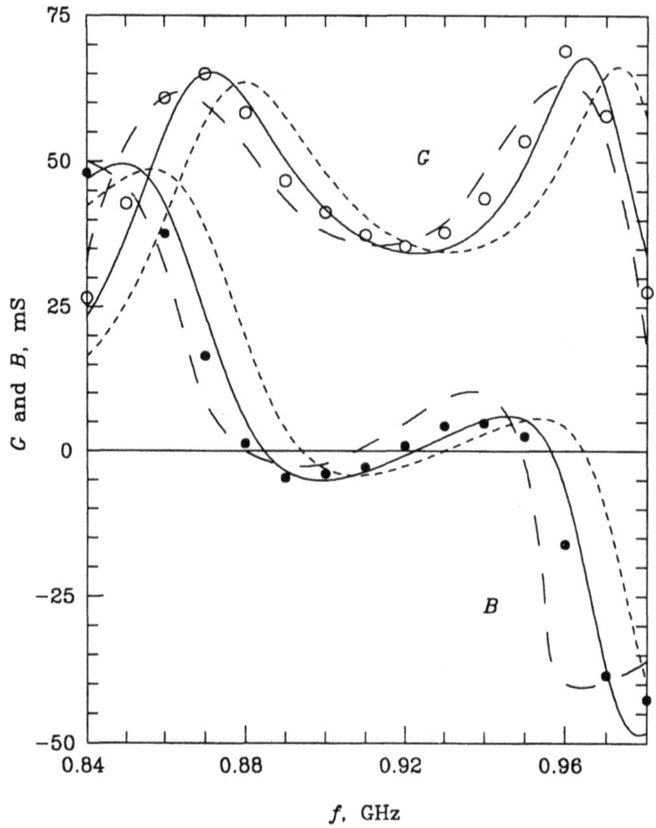

Figure 8.6 Conductance G and susceptance B of the Yagi–Uda array sketched in Figure 8.5 against frequency

$h_1 = 73.3$ mm, $h_2 = h_3 = 66.5$ mm, $h_4 = 64.3$ mm, $h_5 = 79.0$ mm, $d_1 = 43$ mm, $d_2 = 86$ mm, $d_3 = 129$ mm, $d_4 = 63$ mm, $a = 3$ mm, $b/a = 2.3$
—— this method, 15 unknowns, with end effect
.... this method, 10 unknowns, without end effect
– – theory, 10 unknowns [33]
○, ● experiment [33]

for $h = h_1$. It is seen that even with a much smaller correction to the antenna height than for the method from Reference 96, slightly better agreement is achieved with the experiment.

8.3 Surface scatterers

To illustrate some possibilities of the method relating to surface structures, in this Section three types of surface scatterer are analysed. In Section 8.3.1 a square-plate scatter is considered, in Section 8.3.2 a corner scatterer, and in Section 8.3.3 a scatterer in the form of a metal cube.

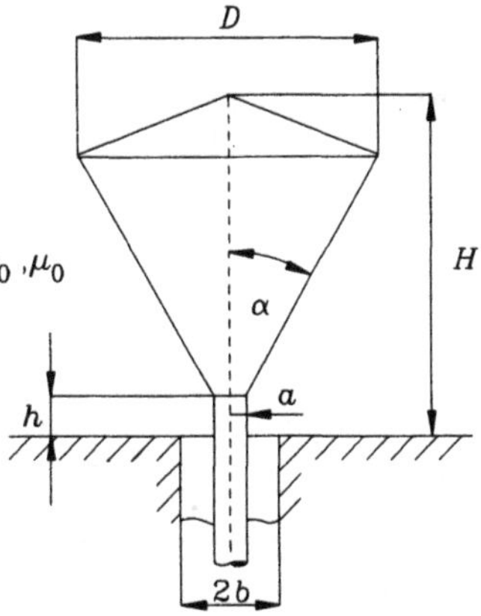

Figure 8.7 Sketch of a conical monopole antenna above the ground plane

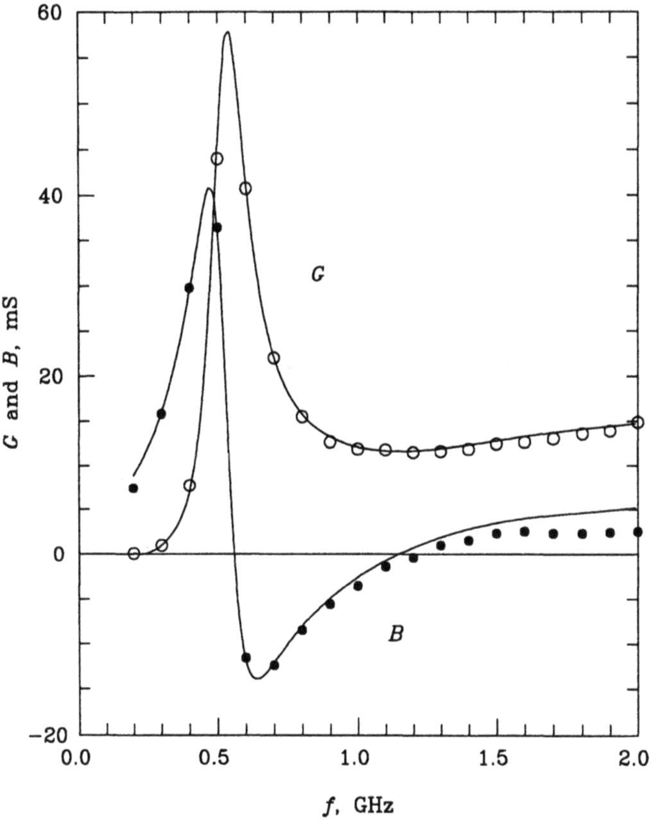

Figure 8.8 Conductance G and susceptance B of the antenna sketched in Figure 8.7 against frequency

h = 1.5 mm, H = 88.6 mm, D = 80 mm, $\alpha = 30°$, a = 3 mm, b = 6.9 mm
—— this method, six unknowns
○, ● experiment [77]

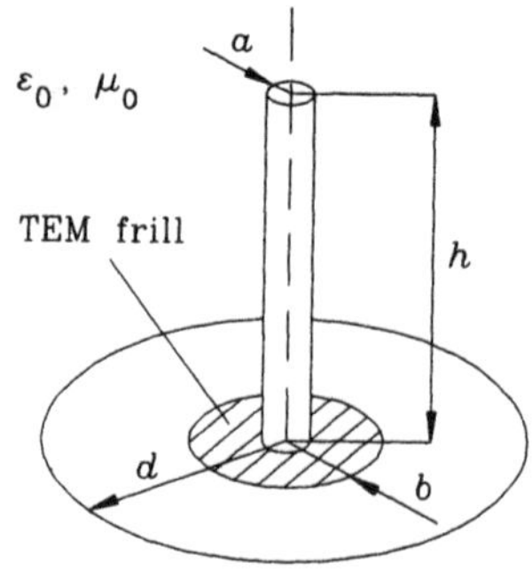

Figure 8.9 *Sketch of a monopole wire antenna above a circular plate*

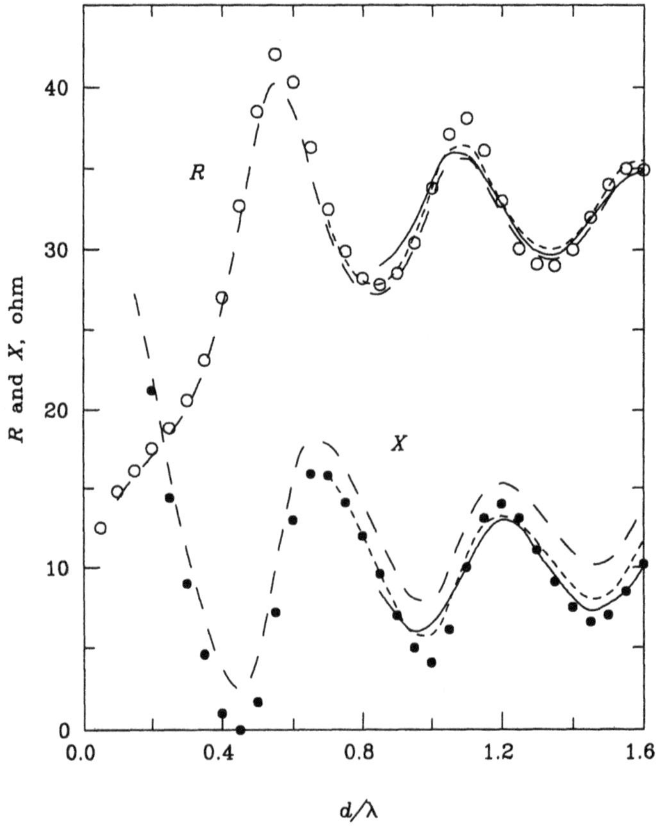

Figure 8.10 *Resistance R and reactance X of the monopole antenna sketched in Figure 8.9 against d/λ*

$h = h_0$, h_1 or h_2, $h_0 = 0.224\,\lambda$, $h_1 = 0.229\,\lambda$, $h_2 = 0.2255\,\lambda$,
$a = 0.003\,\lambda$, $b/a = 2.3$
— — this method, six unknowns $h = h_0$
.... this method, six unknowns, $h = h_2$
○, ● theory, 14 unknowns, $h = h_1$ [96]
—— experiment, $h = h_0$ [96]

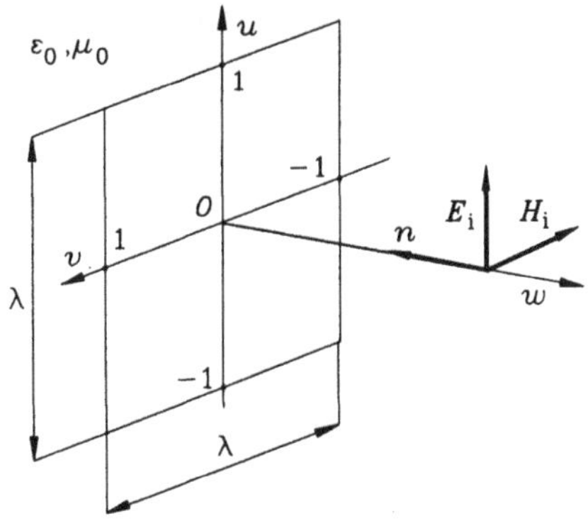

Figure 8.11 Sketch of a square scatterer

8.3.1 Square-plate scatterer

In this Section a square scatterer in a vacuum, of size $\lambda \times \lambda$, is analysed, as shown in Figure 8.11. The scatterer is excited by a uniform plane wave incident normally on the scatterer surface. The incident wave is polarised in the direction of the u-axis.

The analysis of the scatterer was performed starting from the EFIE, and was solved by the Galerkin method and by the point-matching procedure. The total number of unknowns amounted to $\mathcal{N} = 8$.

Figure 8.12 shows the modulus of the u component of the surface-current-density vector along the v axis, normalised with respect to the incident magnetic field, i.e. the ratio $|J_{su}(v, 0)/H_i|$. The theoretical results from Reference 47 are also shown, obtained using a current approximation with $\mathcal{N} = 162$ unknowns, as well as those from Reference 49, obtained with $\mathcal{N} = 113$ unknowns. (In neither of these papers was symmetry taken into account.) Good agreement of all the theoretical results can be observed.

8.3.2 Corner scatterer

Sketched in Fig. 8.13 is a scatterer in the form of two rectangular plates of different sizes, interconnected along a side at an angle. We consider this structure as a scatterer in the electromagnetic field of a plane wave incident normal to the larger plate, polarised parallel to the u axis, and assume that $a = 2/3\lambda$, $b = 1/3\lambda$, $h = \lambda$ and $\alpha = 50°$. The total number of unknowns amounted to $\mathcal{N} = 34$.

Figure 8.14 shows the modulus of the u component of the surface current density along the v axis of this asymmetrical corner scatterer, normalised with respect to the intensity of the incident magnetic field. The theoretical results obtained by the proposed method, using $\mathcal{N} = 34$ unknowns, are compared with the theoretical results from Reference 49 ($\mathcal{N} = 96$ unknowns) and from Reference 47 ($\mathcal{N} = 162$ unknowns). Reasonable agreement is seen between the

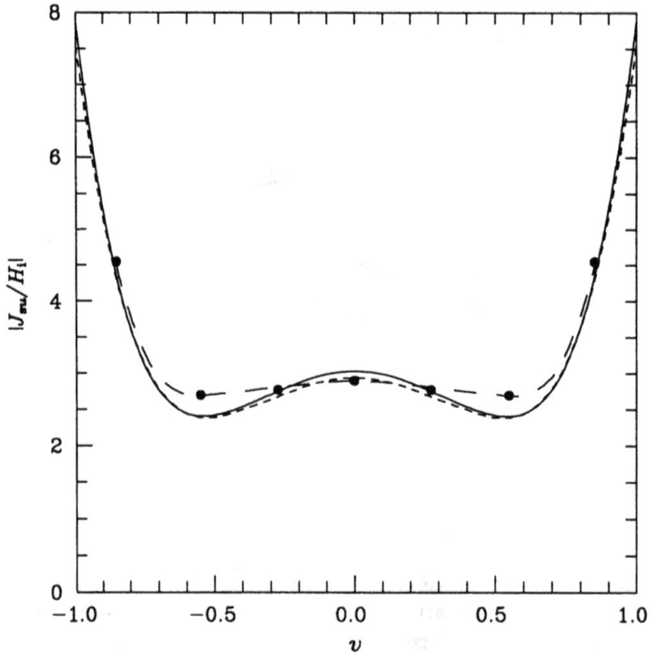

Figure 8.12 *Normalised modulus $|J_{su}(v, 0)/H_i|$ of the u component of the surface-current-density vector along the v axis on the square scatterer sketched in Figure 8.11, of side length λ*

— — this method, eight unknowns
—— point-matching, equidistant matching points
· · · · theory [49]
● ● ● theory [47]

results obtained by the present method and those from Reference 47 while larger discrepancies are observed if these results are compared with those from Reference 49.

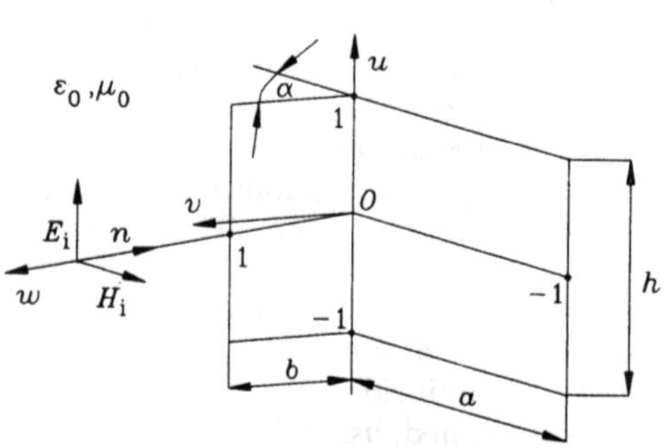

Figure 8.13 *Sketch of an asymmetrical corner scatterer*

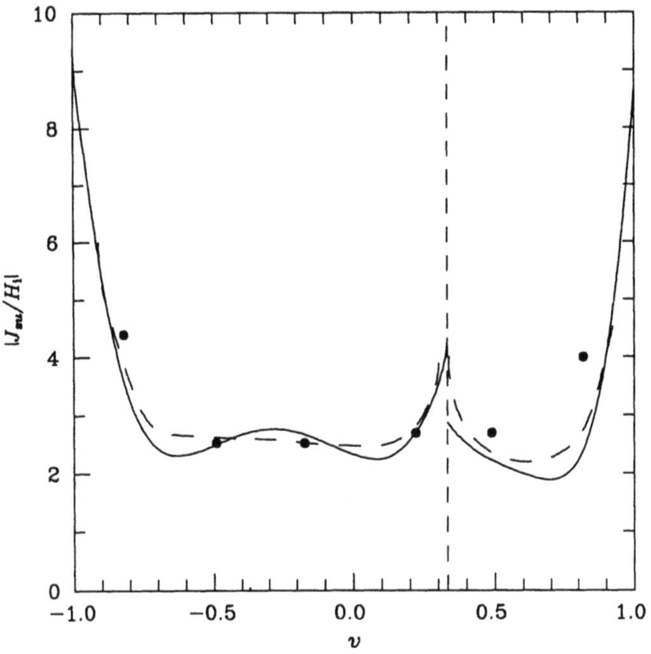

Figure 8.14 *Normalised modulus $|J_{su}(v, 0)/H_i|$ of the u component of the surface-current-density vector along the v axis on the asymmetrical corner scatterer sketched in Figure 8.13*

$a = 2/3\,\lambda, b = 1/3\,\lambda, h = \lambda$ and $\alpha = 50°$
—— this method, 34 unknowns
– – theory, 162 unknowns [47]
● ● ● theory, 96 unknowns [49]

8.3.3 Radar cross-section of a cube

As the final example of a scatterer, consider a metal cube excited as indicated in Figure 8.15. Figure 8.16 shows the monostatic (radar) cross-section of this scatterer, plotted against $4a/\lambda$, obtained by analysing the structure by means of the CFIE. The total number of unknowns amounted to $N = 96$. Also shown in the Figure are experimental results from Reference 97, indicating good agreement of the theoretical results with experiment. Note that for the largest electrical length of the side considered the area of the cube amounts to about $25\lambda^2$.

8.4 Combined wire-and-plate antennas

The aim of this book was to develop an efficient and accurate method for the analysis of arbitrary metallic antennas and scatterers. In many cases such structures are combinations of wires and plates. This Section is devoted to the

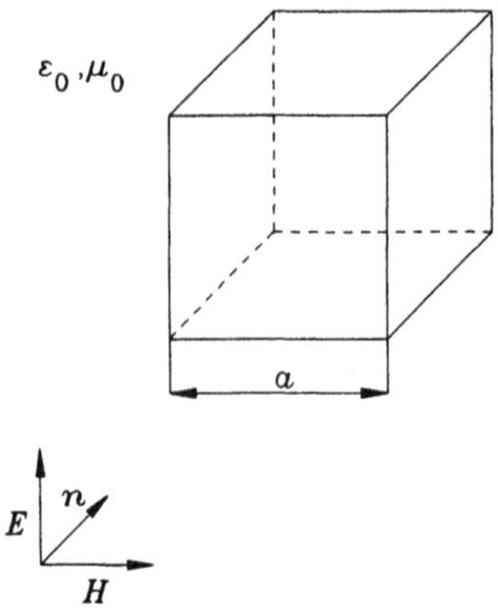

Figure 8.15 Metal cube excited by a plane wave

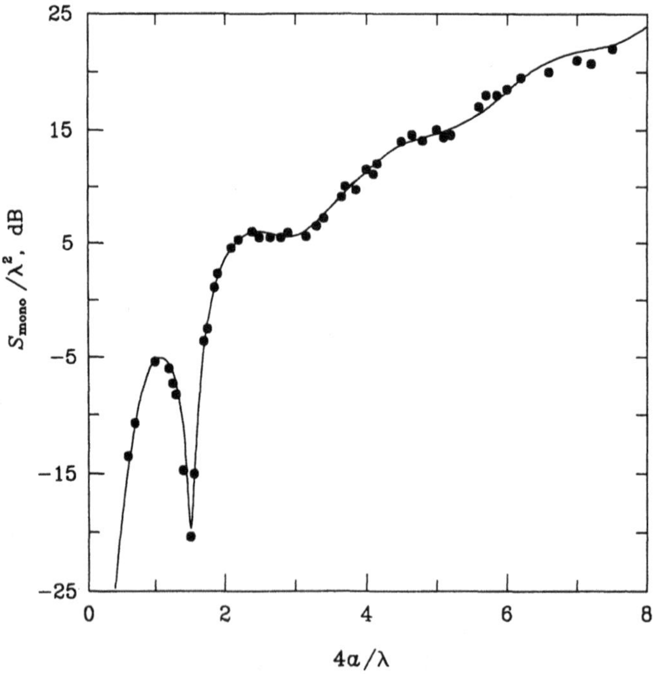

Figure 8.16 Monostatic (radar) cross-section of the cube sketched in Figure 8.15, against $4a/\lambda$

—— this method
● ● ● experiment [97]

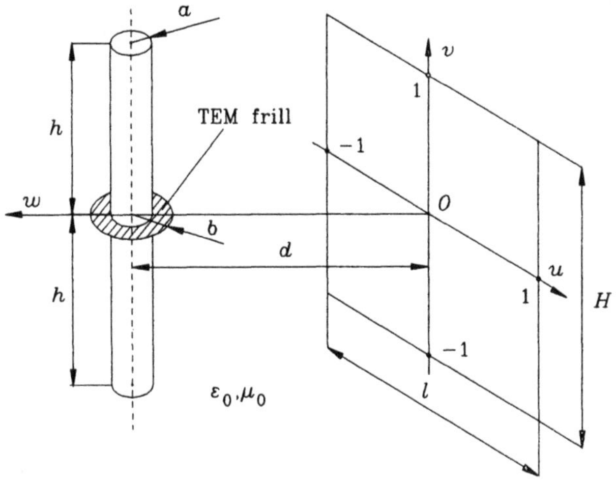

Figure 8.17 Dipole antenna with rectangular reflector

application of the present method to such cases. In Sections 8.4.1 and 8.4.2, a monopole antenna with a plate reflector is considered, and in Section 8.4.3 a planar triangular-plate antenna is analysed. A circular microstrip antenna is considered in Section 8.4.4, and one of rectangular shape in Section 8.4.5. A monopole wire antenna with a parasitic element mounted on a square plate is analysed in Section 8.4.6. In Section 8.4.7 results are presented for a monopole wire antenna mounted on a metal cube which, in turn, is lying on a ground plane.

8.4.1 Monopole wire antenna with rectangular reflector

In this Section a monopole antenna with a rectangular reflector driven by a coaxial line in the monopole ground plane is analysed. The image theory enables this system to be considered approximately as that of a dipole antenna with a rectangular reflector (Figure 8.17).

Figure 8.18 shows the antenna conductance and susceptance, for $h = 126\,\text{mm}$, $a = 3\,\text{mm}$, $b = 6.9\,\text{mm}$, $d = 58.4\,\text{mm}$, $l = 247\,\text{mm}$ and $H = 362\,\text{mm}$, plotted against frequency. If symmetry is taken into account, this results in $N = 15$ unknowns. Shown also in the Figures are the theoretical results from Reference 50 obtained with $N = 14$ unknowns, and the experimental results from the same paper. Good agreement can be observed between the theoretical results obtained by the method developed in this book and experimental results, significantly better than the agreement with the other theoretical results.

Table 8.1 shows the antenna admittance ($h = 75\,\text{mm}$, $a = 3\,\text{mm}$, $b = 6.9\,\text{mm}$, $d = 90\,\text{mm}$, $l = 75\,\text{mm}$ and $H = 180\,\text{mm}$) and its forward and backward gain, obtained by the present method for different degrees of the polynomial approximation for current. For comparison, the results from

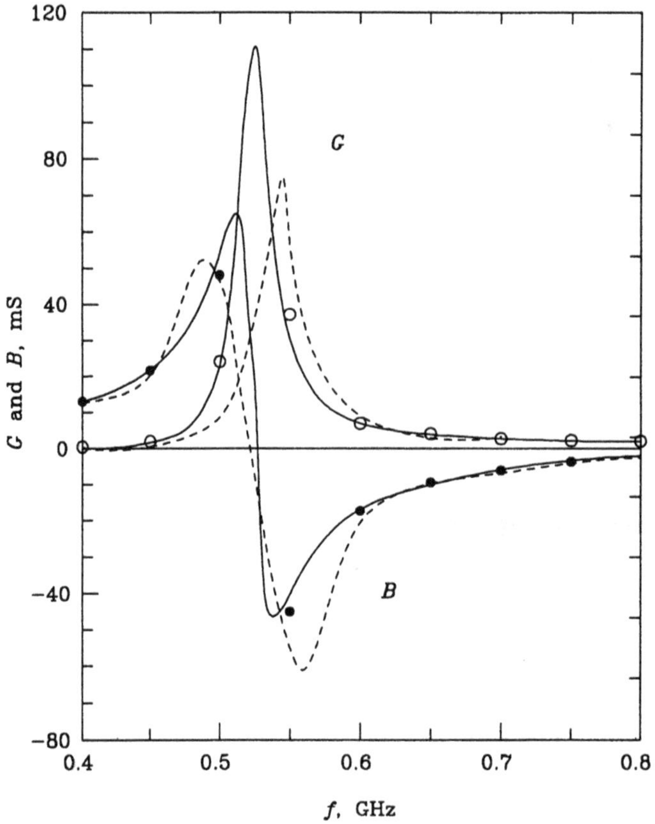

Figure 8.18 *Conductance G and susceptance B of the antenna sketched in Figure 8.17 against frequency*

$h = 126$ mm, $a = 3$ mm, $b = 6.9$ mm, $d = 58.4$ mm, $l = 247$ mm, $H = 362$ mm
—— this method, 15 unknowns
- - - theory, 14 unknowns [50]
○, ● experiment [50]

Reference 50 are shown in Table 8.2, except that in Reference 50 the dipole arm is divided into two subsegments, with polynomial degrees in current approximation n_1 and n_2. It is seen that the present methods yields very stable results, much more stable than those from Reference 50. This is probably because of the application of the Galerkin method, as against point-matching in Reference 50.

8.4.2 Monopole wire antenna with corner reflector

Figure 8.19 shows a sketch of the symmetrical equivalent of a monopole antenna with a corner reflector. Shown in Figure 8.20 is the conductance and susceptance of this antenna ($h = 100$ mm, $a = 3$ mm, $b = 6.9$ mm, $d = 100$ mm, $\alpha = 90°$,

Table 8.1 Admittance and forward and backward gain of the antenna sketched in Fig. 8.17

n_1	n_u	n_v	Y (mS)	$d(0)$ (dB)	$d(\pi)$ (dB)
5	3	3	$12.466 - j5.114$	5.403	−3.809
5	3	4	$12.487 - j5.113$	5.403	−3.846
5	3	5	$12.487 - j5.113$	5.403	−3.846
5	4	4	$12.465 - j5.111$	5.426	−3.846
5	5	5	$12.465 - j5.112$	5.425	−3.840
5	6	6	$12.480 - j5.097$	5.420	−3.888
6	4	4	$12.316 - j4.958$	5.447	−3.825
7	4	4	$12.348 - j5.007$	5.428	−3.840

$h = 75$ mm, $a = 3$ mm, $b = 6.9$ mm, $d = 90$ mm, $l = 75$ mm, $H = 180$ mm
Obtained using the proposed method, for different degrees of the polynomial approximation of current
n_1 = degree of the polynomial along one dipole arm
n_u = degree of the polynomial along the u axis of the reflector
n_v = degree of the polynomial along the v axis of the reflector

$l = 100$ mm, $H = 300$ mm), plotted against frequency. The total number of unknowns amounted to $N = 24$.

The theoretical results obtained by the present method with $N = 24$ unknowns are compared with the theoretical results (obtained with $N = 23$ unknowns) and the experimental results from Reference 51. Excellent agreement is observed between the results obtained by the present method and the experimental results, while the other theoretical results are seen to be significantly less accurate. The two sets of theoretical results being obtained using similar expansions, this difference in accuracy can be explained only by the use of different methods (Galerkin, against point-matching).

Figure 8.21 shows the normalised directive gain of the dipole antenna with

Table 8.2 Theoretical and experimental results, as in Table 8.1, obtained by the method of Reference 50

n_1	n_2	n_u	n_v	Y (mS)	$d(0)$ (dB)	$d(\pi)$ (dB)
3	3	3	4	$12.24 - j5.66$	5.15	−0.25
4	3	3	4	$11.77 - j5.48$	5.00	−0.40
3	4	3	4	$11.50 - j5.53$	5.09	−0.30
3	3	4	4	$11.91 - j6.63$	4.61	0.71
3	3	3	5	$11.95 - j5.32$	5.07	0.15

n_1 and n_2 are degrees of the polynomials along two segments into which the dipole arm is subdivided

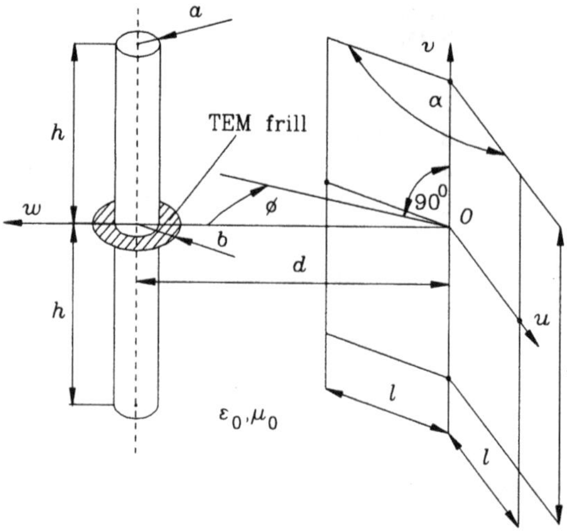

Figure 8.19 Sketch of symmetrical equivalent of a monopole antenna with a corner reflector

a corner reflector ($h = 0.25\lambda$, $a = 0.005\lambda$, $b = 2.3a$, $d = 0.75\lambda$, $\alpha = 120°$, $l = \lambda$, $H = 2.75\lambda$) in the equatorial plane. The normalisation coefficient is $g_d(0°, 90°) = 4.05$ dB. The total number of unknowns amounted to $N = 48$.

The theoretical results obtained by the proposed method are compared with theoretical results (obtained with $N = 61$ unknowns for the u component of the current) and experimental results from Reference 44. Reasonable agreement is observed between the results obtained by the proposed method and the experimental results, while the theoretical results from Reference 44 are visibly less accurate in back radiation.

8.4.3 Triangular plate antenna

Sketched in Figure 8.22 is the symmetrical equivalent of a triangular planar antenna above a ground plane, driven by a coaxial line. Figure 8.23 shows the conductance and susceptance of the antenna ($h = 18$ mm, $H = 108$ mm, $a = 3$ mm, $b = 6.9$ mm, $\alpha = 60°$) plotted against frequency. The antenna was analysed using two different polynomial approximations, resulting in $N = 18$ and $N = 28$ unknowns, respectively.

The theoretical results obtained using the proposed method, with $N = 18$ and $N = 28$ unknowns, are compared in Figure 8.23 with experimental results from Reference 61. Good agreement can be observed between the two sets of results. The theoretical results from Reference 61, obtained with $N = 38$ unknowns, are not plotted, because they follow approximately a curve between the two curves shown.

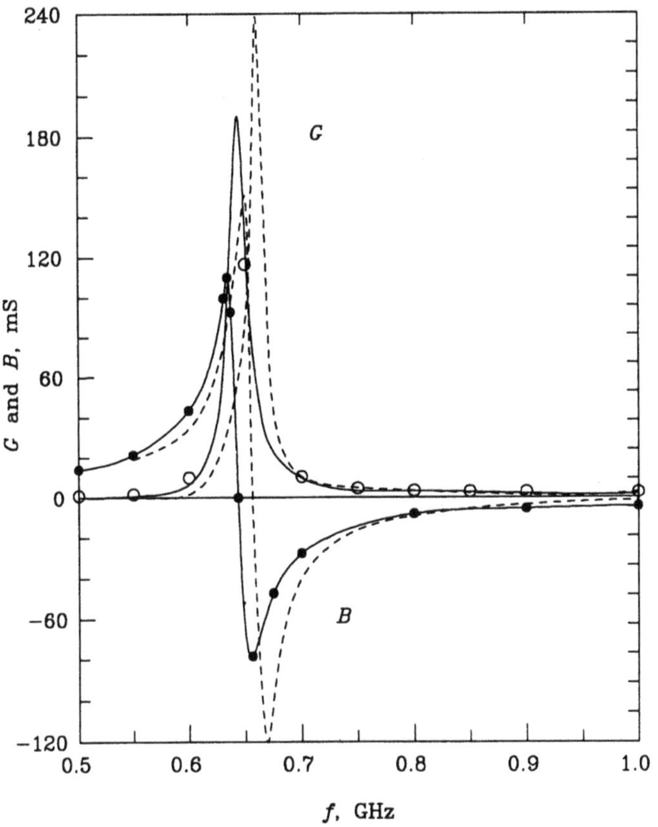

Figure 8.20 *Conductance G and susceptance B of the monopole antenna sketched in Figure 8.19 against frequency*

$h = 100\,\text{mm}$, $a = 3\,\text{mm}$, $b = 6.9\,\text{mm}$, $d = 100\,\text{mm}$, $\alpha = 90°$, $l = 100\,\text{mm}$, $H = 300\,\text{mm}$

—— this method, 24 unknowns

- - - theory, 23 unknowns [51]

○, ● experiment [51]

8.4.4 Circular air-filled microstrip antenna

Sketched in Figure 8.24 is an air-filled circular microstrip antenna. Figure 8.25 shows the antenna conductance and susceptance ($h/R = 0.05$, $a/R = 0.02$, $b/a = 2.3$), plotted against the antenna electrical radius. The results obtained using the present method ($N = 6$ unknowns) are practically identical to the theoretical results from Reference 98 (obtained by taking into account precisely the singularities at the excitation point and at the edges of the circular plates), which are therefore not shown.

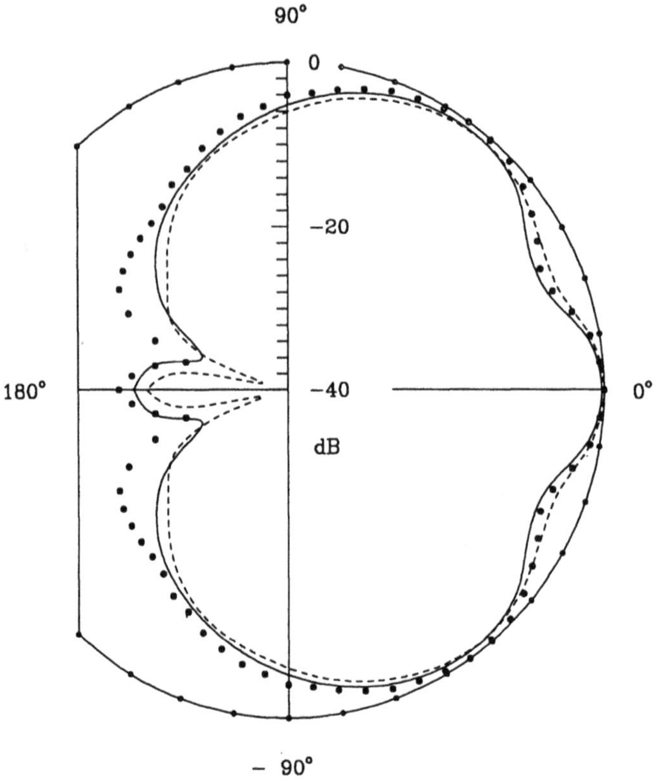

Figure 8.21 Directive gain of dipole antenna with corner reflector sketched in Figure 8.19 in the equatorial plane, normalised with respect to $g_d(0°, 90°) = 4.05$ *dB*

($h = 0.25\,\lambda$, $a = 0.005\,\lambda$, $b = 2.3\,a$, $d = 0.75\,\lambda$, $\alpha = 120°$, $l = \lambda$, $H = 2.75\,\lambda$
—— this method, 48 unknowns
· · · theory, 61 unknowns [44]
- - - - experiment [44]

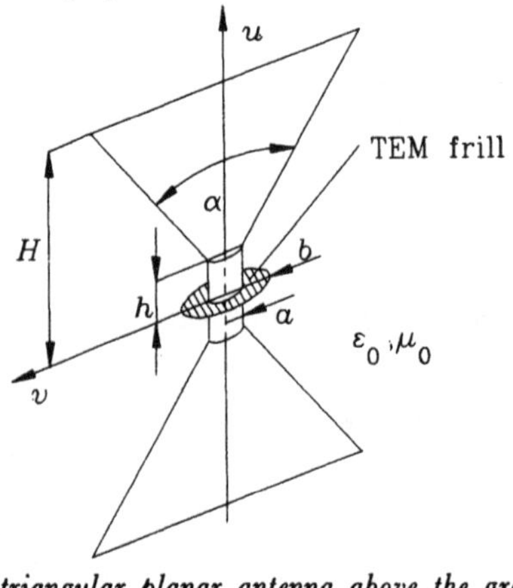

Figure 8.22 Sketch of a triangular planar antenna above the ground plane, driven by a coaxial line

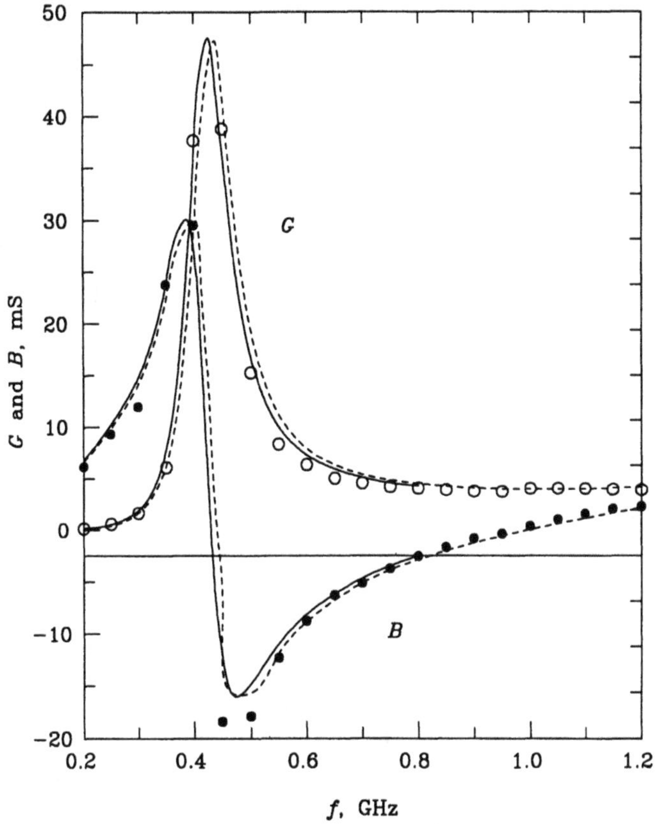

Figure 8.23 *Conductance G and susceptance B of the antenna sketched in Figure 8.22 against frequency*

$h = 18\,\text{mm}$, $H = 108\,\text{mm}$, $a = 3\,\text{mm}$, $b = 6.9\,\text{mm}$, $\alpha = 60°$
—— this theory, 18 unknowns
- - - - this theory, 28 unknowns
○, ● experiment [61]

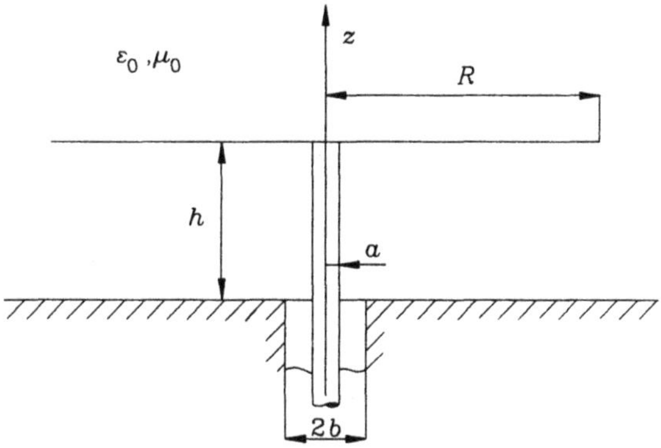

Figure 8.24 *Sketch of an air-filled circular microstrip antenna*

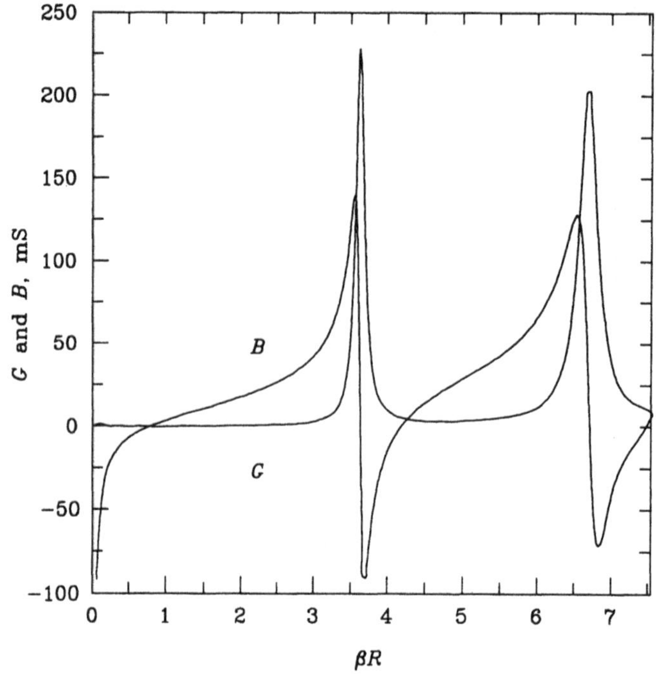

Figure 8.25 *Conductance G and susceptance B of the antenna sketched in Figure 8.24 against the plate electrical radius*

The results were obtained by the present method, with six unknowns $h/R=0.05$, $a/R=0.02$, $b/a=2.3$

8.4.5 Rectangular air-filled microstrip antenna

In this Section results are presented for a rectangular air-filled microstrip antenna, driven by a coaxial line in the ground plane. Image theory yields a symmetrical antenna equivalent, excited by a TEM magnetic-current frill, as sketched in Figure 8.26.

Figure 8.27*a* shows the resistance of this antenna ($h = 6$ mm, $a = 0.65$ mm,

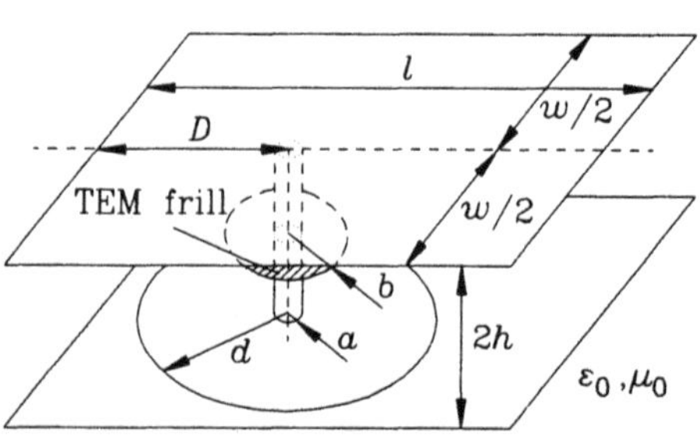

Figure 8.26 *Sketch of symmetrical equivalent of a rectangular air-filled microstrip antenna, driven by coaxial line in the ground plane*

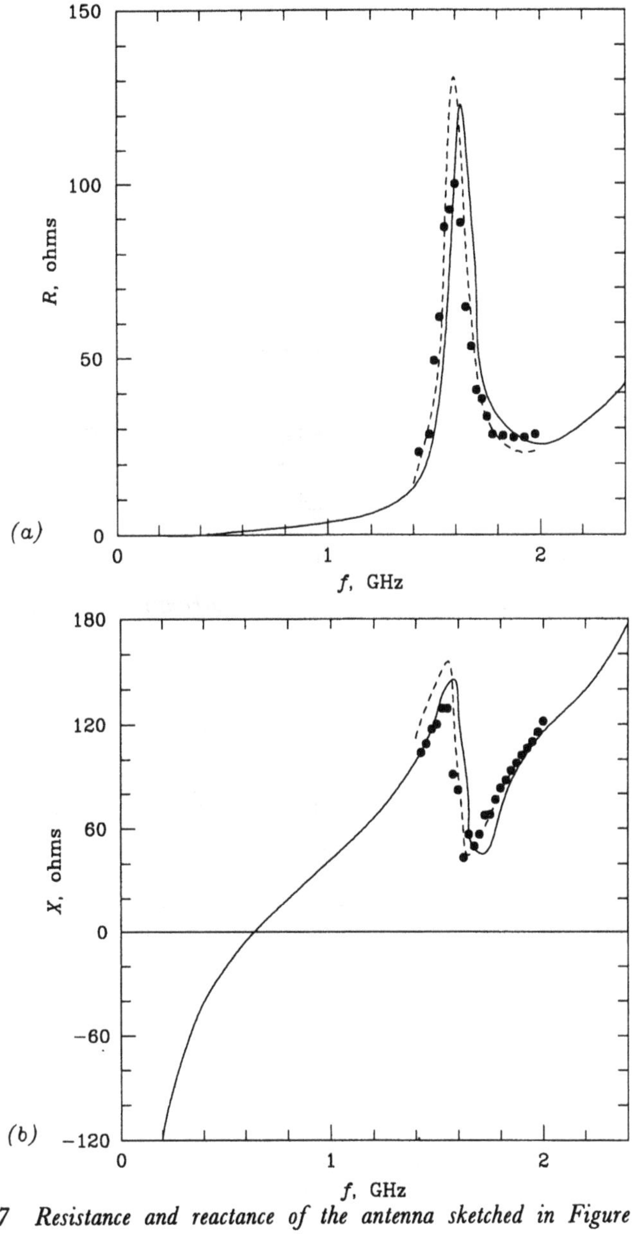

Figure 8.27 *Resistance and reactance of the antenna sketched in Figure 8.26 against frequency*

$h = 6$ mm, $a = 0.65$ mm, $b = 2.05$ mm, $l = 75$ mm, $w = 37.5$ mm, $d = 26.26$ mm
a Resistance
b Reactance
—— this method, 33 unknowns
- - - - theory, 96 unknowns [99]
● ● ● experiment [99]

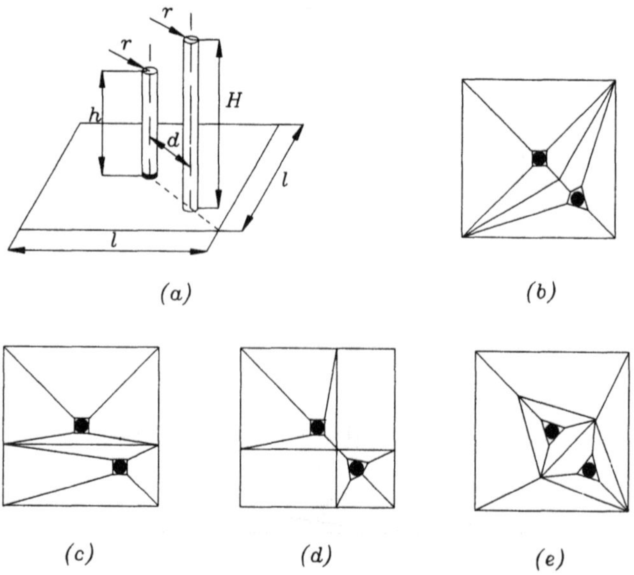

Figure 8.28 *Monopole wire antenna with a parasitic element mounted on a square plate*
a Sketch of the antenna
b–e Examples of partitioning of the plate

$b = 2.05$ mm, $l = 75$ mm, $w = 37.5$ mm and $d = 26.26$ mm), and Fig. 8.27*b* the reactance, in each case plotted against frequency. In this example, modelling of geometry was performed by a cylinder for the inner coaxial-line conductor feeding the antenna, a large flat ring, indicated in the Figure in broken lines (the attachment mode), and a rectangle for the plate. The total number of unknowns amounted to $N = 33$ unknowns.

Also shown in the Figure are the theoretical results (obtained with 96 unknowns) and the experimental results from Reference 99. The theoretical results from Reference 99 were obtained with the same type of attachment mode. Reasonable agreement can be observed between the three sets of results.

8.4.6 *Monopole wire antenna with a parasitic element on a square plate*

Sketched in Figure 8.28 is a monopole wire antenna with a parasitic element mounted on a square plate. Figure 8.29 shows the conductance and susceptance of the antenna, at the plate centre, plotted against the distance d of the parasitic wire element from the monopole, with $h = 421$ mm, $a = 0.8$ mm, $l = 914$ mm and $H = 842$ mm, and at a frequency of $f = 150$ MHz.

The theoretical results obtained by the present method, with $N = 28$ unknowns and $M = 8$ patches, are compared with the theoretical results (obtained with $N = 18$ unknowns) and the experimental results from Reference 46. The results obtained by the present method used the segmentation scheme of the plate sketched in Figure 8.28*b*. Almost the same results were obtained by the segmentation schemes shown in Figures 8.28*c*,

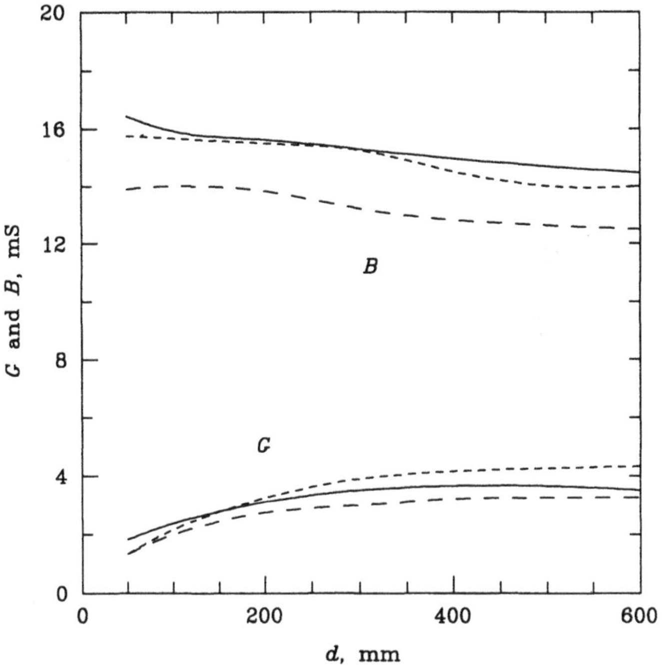

Figure 8.29 *Conductance G and susceptance B of the antenna sketched in Figure 8.28 at the plate centre against the distance of the parasitic wire element from the monopole*

$h = 421$ mm, $a = 0.8$ mm, $l = 914$ mm and $H = 842$ mm, $f = 150$ MHz
—— this method, 28 unknowns
- - - theory, 18 unknowns [46]
· · · · experiment [46]

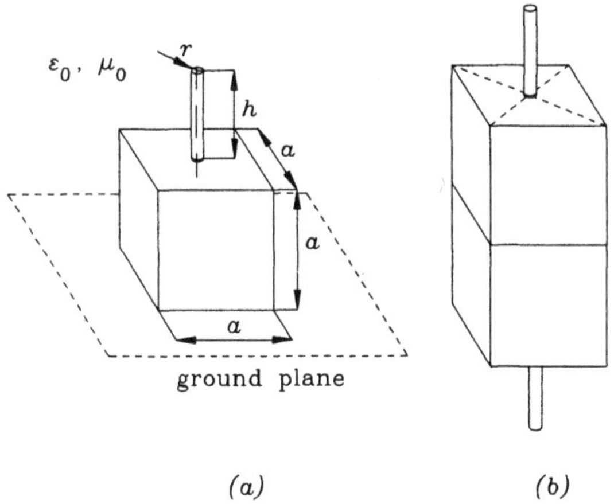

Figure 8.30 *Monopole wire antenna at the centre of a side of a metal cube*

a Monopole wire antenna attached at the centre of a side of metal cube lying on a ground plane
b Symmetrical system to that in (*a*) resulting from image theory

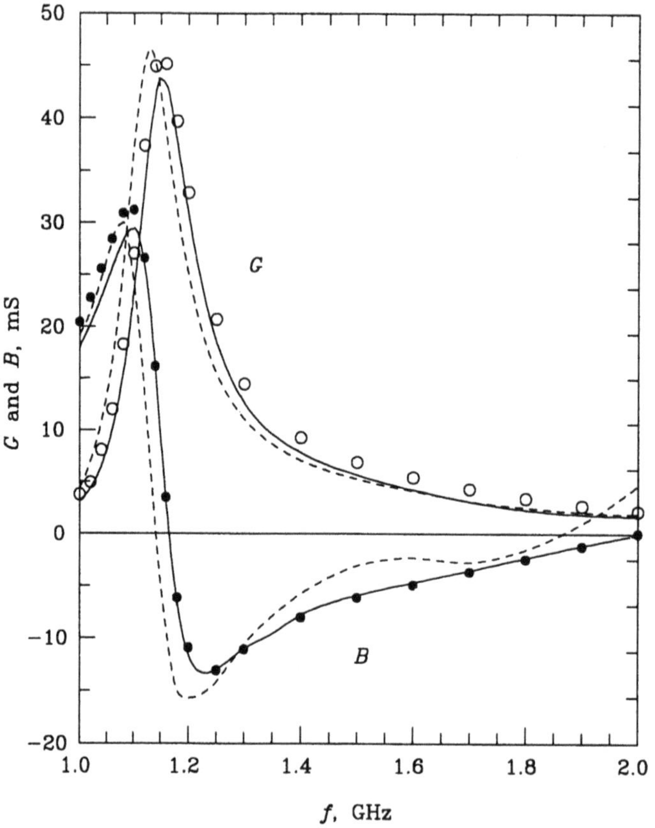

Figure 8.31 *Conductance G and susceptance B of the monopole antenna sketched in Figure 8.30a against frequency*

$a = 100$ mm, $h = 60$ mm
—— this method, 16 patches and 33 unknowns
○, ● theory [83]
- - - - experiment [83]

d and *e*. The theoretical results from Reference 46 were obtained by using the basic attachment modes (one for the driven monopole, and the other for the parasitic element). Very good agreement can be observed between the results obtained by the proposed method and experimental results, while the results for reactance obtained by using the basic attachment mode are seen to be less accurate.

8.4.7 Monopole wire antenna at the centre of the side of a cube lying on a ground plane

As the final example, consider a monopole wire antenna with the base at the centre of a side of a metal cube lying on a ground plane, as sketched in Figure 8.30*a*.

Figure 8.31 shows the antenna conductance and susceptance ($a = 100$ mm,

$h = 60\,\text{mm}$) plotted against frequency. In the first step, image theory was applied to obtain the structure indicated in Figure 8.30*b*. The theoretical results obtained by the present method, with $N = 33$ unknowns and $M = 16$ patches, are compared with the theoretical and the experimental results from Reference 83. The theoretical results from Reference 83 were obtained by using surface-patch modelling for both wires and plates. Satisfactory agreement between theoretical and experimental results is observed.

References

1 KANTOROVICH, L. V., and KRYLOV, V. I.: 'Approximate methods of higher analysis' (John Wiley & Sons, 1959). Translated from Russian

2 KANTOROVICH, L. V., and AKILOV, G. P.: 'Functional analysis in normed spaces' (Pergamon Press, 1964). Translated from Russian

3 NOMURA, Y., and HATTA, T.: 'The theory of a linear antenna, I', *Tech. Rep. Tohoku Univ.*, 1952, **17**, (1)

4 STORM, B.: 'Investigation into modern aerial theory and a new solution of Hallén's integral equation for a cylindrical aerial'. Dissertation, Imperial College, London, 1953

5 HARRINGTON, R. F.: 'Matrix methods for field problems', *Proc. IEEE,* 1967, **55**, (2), pp. 136–149

6 HARRINGTON, R. F.: 'Field computation by moment methods' (McMillan, 1968)

7 POPOVIĆ, B. D.: 'Electromagnetic field theorems (a review)' *Proc. IEE A,* 1981, **128**, pp. 47–63

8 STAKGOLD, I.: 'Boundary value problems of mathematical physics – Vol. 1' (Macmillan, 1970)

9 POCKLINGTON, H. E.: 'Electrical oscillations in wires', *Camb. Phil. Soc. Proc.*, 1897, **9**, pp. 324–332

10 HALLÉN, E.: 'Theoretical investigation into the transmitting and receiving qualities of antennae', *Nova Acta Soc. Sci. Upsal.*, 1938, pp. 1–44

11 KING, R. W. P.: 'The theory of linear antennas' (Harvard University Press, 1956)

12 MEI, K. K.: 'On the integral equations of thin wire antennas', *IEEE Trans.*, 1965, **AP-13**, pp. 374–378

13 RICHMOND, J. H.: 'Digital solutions of the rigorous equations for scattering problems', *Proc. IEEE*, 1965, **53**, pp. 796–804

14 RICHMOND, J. H.: 'Computer analysis of three dimensional wire antennas'. Ohio State Univ. ElectroScience Laboratory, Department of Electrical Engineering, Report 2708-4, 1969

15 OTTO, D. V., and RICHMOND, J. H.: 'Rigorous field expression for piecewise-sinusoidal line source', *IEEE Trans.*, 1969, **AP–17**, p. 98

16 POPOVIĆ, B. D.: 'Polynomial approximation of current along thin symmetrical cylindrical dipoles', *Proc. IEE*, 1970, **117**, pp. 873–878

17 POPOVIĆ, B. D.: 'On polynomial approximation of current along thin asymmetrical cylindrical dipoles', *IEEE Trans.*, 1971, **AP–19**, pp. 117–120

18 TAYLOR, C. D.: 'Electromagnetic scattering from arbitrary configurations of wires', *IEEE Trans.*, 1969, **AP–17**, pp. 662–663

19 TAYLOR, C. D., LIN, S. M., and MCADAMS, H. V.: 'Scattering from crossed wires', *IEEE Trans.*, 1970, **AP–18**, pp. 133–136

20 CHAO, H. H., STRAIT, B. J., and TAYLOR, C. D.: 'Radiation and scattering by configurations of bent wires with junctions', *IEEE Trans.*, 1971, **AP–19**, pp. 701–702

21 SILVESTER, P., and CHAN, K. K.: 'Bubnov–Galerkin solutions to wire antenna problems', *Proc. IEE*, 1972, **119**, pp. 1095–1099

22 SILVESTER, P., and CHAN, K. K.: 'Analysis of antenna structures assembled from arbitrarily located straight wires', *Proc. IEE*, 1973, **120**, pp. 21–26

23 TSAI, L. L.: 'A numerical solution for near and far fields of an annular ring of magnetic current', *IEEE Trans.*, 1972, **AP–20**, pp. 569–576

24 POPOVIĆ, B. D.: 'Thin monopole antenna: finite-size belt-generator representation of coaxial-line excitation', *Proc. IEE*, 1973, **120**, 544–550

25 POPOVIĆ, B. D.: 'Theory of cylindrical antennas with lumped impedance loadings', *Radio Electron. Eng.*, 1973, **43**, pp. 243–248
26 THIELE, G. A.: 'Wire antennas' *in* MITTRA, R. (Ed.): 'Computer techniques for electromagnetics' (Pergamon Press, 1973)
27 IMBRIALE, W. A.: 'Application of the method of moments to thin-wire elements and arrays' *in* MITTRA, R. (Ed.): 'Numerical and asymptotic techniques in electromagnetics' (Springer, 1975)
28 MILLER, E. K., and DEADRICK, F. J.: 'Some computational aspects of thin-wire modelling' *in* MITTRA, R. (Ed.): 'Numerical and asymptotic techniques in electromagnetics' (Springer, 1975)
29 POGGIO, A. J., and MILLER, E. K.: 'Integral equation solutions of three-dimensional scattering problems' *in* MITTRA, R. (Ed.): 'Computer techniques for electromagnetics' (Pergamon Press, 1973)
30 KOMINAMI, M., and ROKUSHIMA, K.: 'A simplified integral equation for thin-wire antennas or scatterers', *Trans. Inst. Electron. Commun. Eng. Jpn. B*, 1976, **59**, pp. 267–273
31 DJORDJEVIĆ, A. R., POPOVIĆ, B. D., and DRAGOVIĆ, M. B.: 'A rapid method for analysis of wire-antenna structures', *Arch. Elektrotech.*, 1979, **61**, pp. 17–23
32 NAKANO, H.: 'The integral equations for a system composed of many arbitrarily bent wires', *Trans. Inst. Electron. Commun. Eng. Jpn. E*, 1982, **65**, pp. 303–309
33 POPOVIĆ, B. D., DRAGOVIĆ, M. B., and DJORDJEVIĆ, A. R.: 'Analysis and synthesis of wire antennas' (Research Studies Press – John Wiley & Sons, 1982)
34 POPOVIĆ, B. D., and NEŠIĆ, A.: 'Generalisation of the concept of equivalent radius of thin cylindrical antennas', *Proc. IEEE H*, 1984, **131**, pp. 153–158
35 POPOVIĆ, B. D., and NEŠIĆ, A.: 'Some extensions of the concept of complementary electromagnetic structures', *Proc. IEE H*, 1985, **132**, pp. 131–137
36 POPOVIĆ, B. D.: 'WireZeus, general program for analysis of wire antennas and scatterers', Distributed by Antenna Software Ltd., Great Malvern, England, 1986
37 POPOVIĆ, B. D.: 'CAD of wire antennas and related radiating structures' (Research Studies Press – John Wiley & Sons, 1991)
38 DJORDJEVIĆ, A. R., BAŽDAR, M. B., VITOŠEVIĆ, G. M., SARKAR, T. K., and HARRINGTON, R. F.: 'Analysis of wire antennas and scatterers – software and user's manual' (Artech House, 1990)
39 OSHIRO, F. K.: 'Source distribution technique for the solution of general electromagnetic scattering problems'. Proceedings of first GISAT Symposium, Vol. 1, Mitre Corporation, 1965, pp. 83–107
40 MAUTZ, J. R., and HARRINGTON, R. F.: 'Radiation and scattering from bodies of revolution', *J. Appl. Sci. Res.*, 1969, **20**, pp. 405–413
41 KNEPP, D. L., and GOLDHIRSH, J.: 'Numerical analysis of electromagnetic radiation properties of smooth conducting bodies of arbitrary shape', *IEEE Trans.*, 1972, **AP–20**, pp. 383–388
42 ALBERTSEN, N. C., HANSEN, J. E., and JENSEN, N. E.: 'Computation of radiation from wire antennas on conducting bodies', *IEEE Trans.*, 1974, **AP–22**, pp. 200–206
43 BURKE, G. J., and POGGIO, A. J.: 'Numerical electromagnetic code – Method of moments'. Technical document 116, AFWL-TR-76-320, Naval Ocean Systems Center, 1977
44 WANG, N. N., RICHMOND, J. H., and GILREATH, M. C.: 'Sinusoidal reaction formulation for radiation and scattering from conducting surfaces', *IEEE Trans.*, 1975, **AP–23**, pp. 376–382
45 RICHMOND, J. H., POZAR, D. M., and NEWMAN, E. H.: 'Rigorous near-zone field expressions for rectangular sinusoidal monopole', *IEEE Trans.*, 1978, **AP–26**, pp. 509–510
46 NEWMAN, E. H., and POZAR, D. M.: 'Electromagnetic modeling of composite wire and surface geometries', *IEEE Trans.*, 1978, **AP–26**, pp. 784–789
47 GLISSON, A. W.: 'On the development of numerical techniques for treating arbitrarily-shaped surfaces'. PhD dissertation, University of Mississippi, 1978
48 GLISSON, A. W., and WILTON, D. R.: 'Simple and efficient numerical methods for problems of electromagnetic radiation and scattering from surfaces', *IEEE Trans.*, 1980, **AP–28**, pp. 593–603

49 RAO, S. M., WILTON, D. R., and GLISSON, A. W.: 'Electromagnetic scattering by surfaces of arbitrary shape', *IEEE Trans.*, 1982, **AP–30**, pp. 409–418
50 POPOVIĆ, B. D., and DJORDJEVIĆ, A. R.: 'Analysis of monopole antenna with rectangular reflector'. Proceedings of 2nd ICAP, York, *IEE Conf. Publ. 195*, 1981, pp. 513–517
51 KOLUNDŽIJA, B. M., and DJORDJEVIĆ, A. .R.: 'Analysis of dipole antenna with corner reflector'. Proceedings of 7th colloquium on microwave communication, Budapest, 1982, pp. 319–322
52 SHAEFFER, J. F., and MEDGYESI–MITSCHANG, L. N.: 'Radiation from wire antenna attached to bodies of revolution: the junction problem', *IEEE Trans.*, 1981, **AP–29**, pp. 479–487
53 SHAEFFER, J. F.: 'EM scattering from bodies of revoluton with attached wires', *IEEE Trans.*, 1982, **AP–30**, pp. 426–431
54 MEDGYESI-MITSCHANG, L. N., and EFTIMIU, C.: 'Scattering from wires and open circular cylinders of finite length using entire domain Galerkin expansions', *IEEE Trans.*, 1982, **AP–30**, pp. 628–636
55 MEDGYESI-MITSCHANG, L. N., and PUTNAM, J. M.: 'Scattering from finite bodies of translation: plates, curved surfaces, and noncircular cylinders', *IEEE Trans.*, 1983, **AP–31**, pp. 847–852
56 MEDGYESI-MITSCHANG, L. N., and PUTNAM, J. M.: 'Formulation for wire radiators on bodies of translation with and without end caps', *IEEE Trans.*, 1983, **AP–31**, pp. 853–862
57 NEWMAN, E. H., and TULYATHAN, P.: 'A surface patch model for polygonal plates', *IEEE Trans.*, 1982, **AP–30**, pp. 588–593
58 POZAR, D. M., and NEWMAN, E. H.: 'Analysis of a monopole mounted near or at the edge of a half-plane', *IEEE Trans.*, 1982, **AP–29**, pp. 488–495
59 POZAR, D. M., and NEWMAN, E. H.: 'Analysis of a monopole mounted near an edge or a vertex', *IEEE Trans.*, 1982, **AP–30**, pp. 401–408
60 NEWMAN, E. H., and ALEXANDROPULOS, M.: 'Polygonal plate modeling of realistic structures', *IEEE Trans.*, 1984, **AP–32**, pp. 742–747
61 POPOVIĆ, B. D., and KOLUNDŽIJA, B. M.: 'Analysis of a class of symmetrical thin-plate triangular antennas', *Proc. IEE H*, 1987, **134**, pp. 205–210
62 KOLUNDŽIJA, B. M., and POPOVIĆ, B. D.: 'Analysis and synthesis of a class of broadband symmetrical planar antennas'. Proceedings of URSI symposium, Budapest, 1986, Pt. A, pp. 669–671
63 BORNHOLDT, J. M., and MEDGYESI-MITSCHANG, L. N.: 'Mixed-domain Galerkin expansions in scattering problems', *IEEE Trans.*, 1988, **AP–36**, pp. 216–227
64 KOLUNDŽIJA, B. .M.: 'Electromagnetic modeling of wire-to-plate structures'. DSc Dissertation, University of Belgrade, 1990 (in Serbian)
65 KOLUNDŽIJA, B. M.: 'General entire-domain Galerkin method for electromagnetic modeling of composite wire-to-plate structures'. Proceedings of 20th EuMc, Budapest, 1990, Pt. 1, pp. 853–858
66 KOLUNDŽIJA, B. M., and POPOVIĆ, B. D.: 'Entire-domain Galerkin method for analysis of generalised wire antennas and scatterers', *Proc. IEE H*, 1992, **139**, pp. 17–24
67 KOLUNDŽIJA, B. M., and POPOVIĆ, B. D.: 'Entire-domain Galerkin method for analysis of metallic antennas and scatterers', *Proc. IEE H*, 1993, **140**, pp. 1–10
68 MITTRA, R., and KLEIN, C. A.: 'Stability and convergence of moment solutions' *in* MITTRA, R. (Ed.): 'Numerical and asymptotic techniques in electromagnetics' (Springer, 1975)
69 SARKAR, T. K.: 'A study of the various methods for computing electromagnetic field utilizing thin-wire integral equations', *Radio Sci.*, 1983, **18**, pp. 29–38
70 SARKAR, T. K.: 'A note on the choice of weighting functions in the method of moments', *IEEE Trans.*, 1985, **AP–33**, pp. 436–441
71 SARKAR, T. K., SIARKIEWICZ, K. R., and STRATTON, R. F.: 'Survey of numerical methods for solution of large systems of linear equations for electromagnetic field problems', *IEEE Trans.*, 1981, **AP–29**, 847–856
72 SARKAR, T. K., DJORDJEVIĆ, A. R., and ARVAS, E.: 'On the choice of expansion and weighting functions in the numerical solution of operator equations', *IEEE Trans.*, 1985, **AP–33**, pp. 988–996

73 DJORDJEVIĆ, A. R., and SARKAR, T. K.: 'A theorem on the moment methods', *IEEE Trans.*, 1987, **AP-35**, pp. 353–355

74 MAUTZ, J. R., and HARRINGTON, R. F.: '*H*-field, *E*-field and combined-field solutions for conducting bodies of revolution', *Arch. Elektr. Uebertrag.*, 1978, **32**, pp. 157–164

75 STRATTON, J. A.: 'Electromagnetic theory' (McGraw-Hill, 1941)

76 TAYLOR, D., and WILTON, D. R.: 'The extended boundary condition solution of the dipole antenna of revolution', *IEEE Trans.*, 1972, **AP-20**, pp. 772–776

77 DJORDJEVIĆ, A. R., POPOVIĆ, B. D., and DRAGOVIĆ, M. B.: 'Analysis of electically thick antennas of revolution'. Proceedings of 3rd ICAP, Norwich, *IEE Conf. Publ. 219*, 1983, pp. 390–394

78 FROEBERG, C.-E.: 'Introduction to numerical analysis' (Addison–Wesley, 1965)

79 BARSKY, B. A.: 'Computer graphics and geometric modeling using beta-spline' (Springer, 1988)

80 KOLUNDŽIJA, B. M.: 'Effect of a wire end in thin-wire analysis'. Proceedings of IEEE AP-S Symposium, Syracuse, 1988, pp. 843–846

81 DRAGOVIĆ, M. B.: 'On the analysis of cylindrical antennas with concentrated loadings'. DSc dissertation, University of Belgrade, 1979 (in Serbian)

82 ANDRES, F. L. H., and JAMBRINA, J. L. F.: 'Simulation of curved surfaces by patch modeling in the study of electromagnetic scattering by resonant size bodies'. Proceedings of 20th EuMc, Budapest, 1990, Pt. 2, pp. 1365–1370

83 BHATTACHARYA, S., LONG, S. A., and WILTON, D. R.: 'The input impedance of a monopole antenna mounted on a cubical conducting box', *IEEE Trans.*, 1987, **AP-35**, pp. 756–762

84 KOLUNDŽIJA, B. M., and POPOVIĆ, B. D.: 'General localised junction model in the analysis of wire-to-plate functions' *Proc. IEE H*, 1994, **141**, pp. 1–7

85 DE SMEDT, R., and VAN BLADEL, J. G.: 'Field singularities at the tip of a metallic cone of arbitrary cross section', *IEEE Trans.*, 1986, **AP-34**, pp. 865–871

86 VAN BLADEL, J.: 'Electromagnetic fields' (McGraw–Hill, 1964)

87 WILTON, D. R., and GOVIND, S.: 'Incorporation of edge conditions in moment method solutions', *IEEE Trans.*, 1977, **AP-25**, pp. 845–850

88 SARKAR, T. K., and RAO, S. M.: 'A simple technique for solving E-field integral equations for conducting bodies at internal resonances', *IEEE Trans.*, 1982, **AP-30**, pp. 1250–1254

89 MURRAY, F. J.: 'The solutions of linear operator equations', *J. Math. Phys.*, 1943, **22**, pp. 148–157

90 MEDGYESI-MITSCHANG, L. N., and WANG, D. S.: 'Hybrid solutions at internal resonances', *IEEE Trans.*, 1985, **AP-33**, pp. 671–674

91 YAGHJIAN, A. D.: 'Augmented electric- and magnetic field integral equations', *Radio Sci.*, 1981, **16**, pp. 987–1001

92 MAUTZ, J. R., and HARRINGTON, R. F.: 'A combined-source solution for radiation and scattering from a perfectly conducting body', *IEEE Trans.*, 1979, **AP-27**, pp. 445–454

93 KOUYOUMJIAN, R. G.: 'The calculation of the echo area of perfectly conducting objects by the variational method'. PhD dissertation, Ohio State University, USA, 1969

94 WILTON, D. R., and BUTLER, C. M.: 'Efficient numerical technique for solving Pocklington's equation and their relationships to others methods', *IEEE Trans.*, 1976, **AP-24**, pp. 83–86

95 DJORDJEVIĆ, A. R.: 'On synthesis of thin wire antennas'. DSc dissertation, University of Belgrade, 1979 (in Serbian)

96 RICHMOND, J. H: 'Monopole antenna on circular disc', *IEEE Trans.*, 1984, **AP-32**, pp. 1282–1287

97 YAGHJIAN, A. D., and McGAHAN, R. V.: 'Broadside radar cross section of the perfectly conducting cube', *IEEE Trans.*, 1985, **AP-33**, pp. 322–329

98 PINHAS, S., SHTRIKMAN, S., and TREVES, D.: 'Moment-method solution of the center-fed microstrip disk antenna invoking feed and edge current singularities', *IEEE Trans.*, 1989, **AP-37**, pp. 1516–1522

99 BROWN, K.: 'Integral equation formulation for the rectangular microstrip antenna'. DSc thesis, Technical University of Denmark, Lyngby, 1989

100 KOLUNDŽIJA, B. M., and POPOVIĆ, B. D.: 'A new, rapid and accurate method for evaluation of potential integrals occurring in thin-wire antenna problems'. Proceedings of 7th ICAP, York, *IEE Conf. Publ. 274*, 1987, pp. 35–38

Appendix 1

Evaluation of line integrals of potentials and field vectors

A1.1 Reduction to canonical form of line integrals of potentials and field vectors due to currents along a truncated cone

We have seen in Chapter 5 that, if the so-called reduced kernel is introduced into the expressions for potentials and field vectors due to currents along a truncated cone, certain line integrals need to be evaluated. If polynomial current distribution is adopted, these integrals have the forms (the expressions in eqns. 5.72 and 5.73)

$$P_i = \int_{u_1}^{u_2} u^{i-1} g(R_a)\,\mathrm{d}u \tag{A1.1}$$

$$Q_i = \int_{u_1}^{u_2} u^{i-1} \frac{1}{R_a} \frac{\mathrm{d}g(R_a)}{\mathrm{d}R_a}\,\mathrm{d}u \tag{A1.2}$$

where the Green function $g(R_a)$ and the average distance R_a are given by

$$g(R_a) = \frac{\exp(-j\beta R_a)}{4\pi R_a} \tag{A1.3}$$

$$R_a = \sqrt{[\{\boldsymbol{r} - \boldsymbol{r}_a(u)\}^2 + a(u)^2]} \tag{A1.4}$$

and $\boldsymbol{r}_a(u)$ and $a(u)$ are the parametric equations of the axis and the radius of the truncated cone given in eqns. 2.20 and 2.21. It is convenient to rewrite these parametric equations as

$$\left.\begin{aligned} \boldsymbol{r}_a(u) &= \boldsymbol{r}_c + \boldsymbol{r}_u u \\ a(u) &= a_c + a_u u \end{aligned}\right\} \tag{A1.5}$$

where the vector coefficients $\boldsymbol{r}_c$ and $\boldsymbol{r}_u$ and the scalar coefficients a_c and a_u are obtained as

$$\boldsymbol{r}_c = \boldsymbol{r}_1 - u_1 \boldsymbol{r}_u \qquad \boldsymbol{r}_u = \frac{\boldsymbol{r}_2 - \boldsymbol{r}_1}{u_2 - u_1} \tag{A1.6}$$

$$a_c = a_1 - u_1 a_u \qquad a_u = \frac{a_2 - a_1}{u_2 - u_1} \tag{A1.7}$$

With this notation, the average distance R_a can be written as

$$R_a = \sqrt{[e_u\{(u-u_0)^2 + d_u^2\}]} \tag{A1.8}$$

where the Lamé coefficients e_u and the coefficients u_0 and d_u are obtained as

$$e_u = \sqrt{(\mathbf{r}_u^2 + a_u^2)} \tag{A1.9}$$

$$\left.\begin{aligned} u_0 &= \frac{(\mathbf{r}-\mathbf{r}_c)\,.\,\mathbf{r}_u - a_c a_u}{e_u^2} \\ d_u^2 &= \frac{(\mathbf{r}-\mathbf{r}_c)^2 + a_c^2}{e_u^2} - u_0^2 \end{aligned}\right\} \tag{A1.10}$$

It is next convenient to introduce the electrical lengths that correspond to the quantities u, u_0, d_u and R_a, i.e.

$$s = \beta e_u u \qquad s_0 = \beta e_u u_0 \qquad d_s = \beta e_u d_u \qquad R_0 = \beta R_a \tag{A1.11}$$

The integrals P_i and Q_i can then be written in the form

$$P_i = \frac{\beta}{4\pi}(\beta e_u)^{-i} p_{i-1} \tag{A1.12}$$

$$Q_i = -\frac{\beta^3}{4\pi}(\beta e_u)^{-i} q_{i-1} \tag{A1.13}$$

$$p_i = \int_{s_1}^{s_2} s^i \frac{\exp(-jR_0)}{R_0}\,ds \tag{A1.14}$$

$$q_i = \int_{s_1}^{s_2} s^i \frac{1+jR_0}{R_0^3}\exp(-jR_0)\,ds \tag{A1.15}$$

$$R_0 = \sqrt{\{(s-s_0)^2 + d_s^2\}} \tag{A1.16}$$

The integrals p_i and q_i can be considered to represent canonical forms of the line integrals of potentials and field vectors.

A1.2 Evaluation of line integrals of potentials and field vectors given in canonical form

In the general case, the line integrals of potentials and field vectors in canonical form are given as

$$P = \int_{s_1}^{s_2} f(s)\frac{\exp(-jR_0)}{R_0}\,ds \tag{A1.17}$$

$$q = \int_{s_1}^{s_2} f(s)\frac{1+jR_0}{R_0^3}\exp(-jR_0)\,ds \tag{A1.18}$$

where $f(s)$ is a slowly varying real function in the integration domain and R_0 is given in eqn. A1.16. The complete imaginary parts of the integrand are also slowly varying functions, so that they can be integrated easily using any numerical-quadrature formula. Let us therefore focus our attention on evaluation of the real part of the above integrals, i.e. on evaluation of the integrals of the general form

$$p_r = \int_{s_1}^{s_2} f(s)\, \frac{\cos(R_0)}{R_0}\, \mathrm{d}s \tag{A1.19}$$

$$q_r = \int_{s_1}^{s_2} f(s)\, \frac{\cos(R_0) + R_0 \sin(R_0)}{R_0^3}\, \mathrm{d}s \tag{A1.20}$$

If the field point is relatively far from the integration segment (the critical distance is between one-tenth and one complete length of the integration segment), these integrals can also be integrated efficiently using an appropriate numerical-quadrature formula.

There are several different methods for more or less efficient evaluation of these integrals when the field point is relatively close to the integration segment. Here we describe a method which permits very efficient evaluation of the integrals with only relatively minor additional analytical effort. For the potential integrals, this method has been described in Reference 100.

To obtain accurate numerical integration, the integrands of both integrals are expanded in a series in the following manner. First the function $f(s)$ is expanded in the Taylor series in s about the point $s = s_0$, and the function $\cos R_0$ in the MacLaurin series in R_0. Next, in the expansion for $f(s)$ every $(s - s_0)^2$ is replaced by $(R_0^2 - d_s^2)$. Obviously the integrands can be written in the form

$$f(s)\, \frac{\cos(R_0)}{R_0} = \sum_{i=0}^{\infty} a_{pi} R_0^{2i-1} + (s - s_0) \sum_{i=0}^{\infty} b_{pi} R_0^{2i-1} \tag{A1.21}$$

$$f(s)\, \frac{\cos(R_0) + R_0 \sin(R_0)}{R_0^3}\, \mathrm{d}s = \sum_{i=0}^{\infty} a_{qi} R_0^{2i-3} + (s - s_0) \sum_{i=0}^{\infty} b_{qi} R_0^{2i-3} \tag{A1.22}$$

where a_{pi}, b_{pi}, a_{qi} and b_{qi} are coefficients which need to be determined. This can be done starting from the expression in eqn. A1.21 multiplied by R_0, and the expression in eqn. A1.22 multiplied by R_0^3. In doing this, we first determine the ith derivative with respect to R_0^2 at $R_0 = 0$ of the left- and right-hand side of the expressions thus obtained. This equation is next split into its real and imaginary parts. Finally, after simple manipulations, for the coefficients a_{pi} and b_{pi} we obtain

$$a_{pi} = \frac{1}{i!}\, \mathrm{Re}\left[\frac{\mathrm{d}^i}{\mathrm{d}(R_0^2)^i} \{ f(s) \cos(R_0) \} |_{R_0 = 0} \right] \tag{A1.23}$$

$$b_p i = \frac{1}{d_s i!}\, \mathrm{Im}\left[\frac{\mathrm{d}^i}{\mathrm{d}(R_0^2)^i} \{ f(s) \cos(R_0) \} |_{R_0 = 0} \right] + \frac{1}{d_s i!} \sum_{j=0}^{i-1} c_{ij}\, b_{pj}$$

$$c_{ij} = \frac{i!}{k!}\, \frac{(2k-3)!!}{2^k d_s^{2k-1}} \qquad k = i - j \tag{A1.24}$$

where the sum does not exist for $i = 0$. [The double factorial in the above expressions is defined as $(2k-3)!! = (2k-3) \times (2k-5) \times \cdots \times 3 \times 1$ for $k > 1$, and $(-1)!! = 1$.] The same expressions are valid for a_{qi} and b_{qi}, except that $f(s)\cos(R_0)$ should be replaced by $f(s)\{\cos(R_0) + R_0 \sin(R_0)\}$.

By substituting the expressions in eqns. A1.21 and A1.22 into eqns. A1.19 and A1.20, the integrals p_r and q_r are expressed in the form of infinite series

$$p_r = \sum_{i=0}^{\infty} a_{pi} A_i + \sum_{i=0}^{\infty} b_{pi} B_i \tag{A1.25}$$

$$q_r = \sum_{i=0}^{\infty} a_{qi} A_{i-1} + \sum_{i=0}^{\infty} b_{qi} B_{i-1} \tag{A1.26}$$

where A_i and B_i are integrals of the form

$$A_i = \int_{s_1}^{s_2} R_0^{2i-1}\, ds \tag{A1.27}$$

$$B_i = \int_{s_1}^{s_2} (s - s_0) R_0^{2i-1}\, ds \tag{A1.28}$$

The integrals A_i and B_i can easily be evaluated analytically, noting that

$$\left.\begin{aligned} A_i &= \frac{2i-1}{2i} d_s^2 A_{i-1} + \frac{s-s_0}{2i-1} R_0^{2i-1}\Big|_{s_1}^{s_2} \qquad i > 0 \\ A_0 &= \ln(s - s_0 + R_0)\big|_{s_1}^{s_2} \\ A_{-1} &= \frac{s-s_0}{d_s^2}\frac{1}{R_0}\Big|_{s_1}^{s_2} \end{aligned}\right\} \tag{A1.29}$$

$$B_i = \frac{R^{2i+1}}{2i+1}\Big|_{s_1}^{s_2} \tag{A1.30}$$

It is evident that the integrals p_r and q_r can be evaluated approximately by summing a finite number of the terms in the series in eqns. A1.25 and A1.26. However, such an evaluation usually requires summation of a relatively large number of terms. Therefore for the evaluation of the integrals we combine analytical and numerical evaluation.

Consider again the expression in eqn. A1.21. Note that the first term in both sums has a pole in the complex s plane at the point

$$s = s_0 + jd_s \tag{A1.31}$$

The other terms of these sums are well behaved functions in the complex plane, but their first derivatives in R_0^2 have a pole at the same point. Generally speaking, for the ith terms of these sums it can be found that their derivatives in R_0^2 including the $(i-1)$th derivative are well behaved functions in the complex plane, while starting from the ith derivative they have a pole at the above point. It is obvious that the higher terms of these sums can be evaluated numerically by simple numerical integration more accurately than can the lower terms.

Bearing all this in mind, the following method is proposed for efficient numerical evaluation of the integrals p_r and q_r: the first few terms of the series in eqns. A1.25 and A1.26 are integrated analytically, while the rest are evaluated numerically. Moreover, if u_0 is between u_1 and u_2, the integration segment should be divided into two subsegments at the point u_0.

This method was found to be extremely efficient. The greatest increase in accuracy is obtained by simultaneously increasing the order of the integration formula (e.g. Gauss-Legendre formula) and the number of terms that are evaluated analytically. Thus, for example, if $f(s) = s^6$, $s_1 = 0$, $s_2 = 1$, $s_0 = 1$ and $d_s = 0.01$, by using the Gauss–Legendre formula of the eighth order and by analytical evaluation of the first four terms in both sums in eqn. A1.25, the value of the potential integral is obtained with a relative error less than 10^{-12}. For details the reader is referred to Reference 100.

Appendix 2

Evaluation of the integrals of potentials and field vectors due to polynomial distribution of current over bilinear surfaces

We have seen in Chapter 5 that, in evaluating the potentials and field vectors due to polynomial approximation of current over bilinear surfaces, the following integrals are encountered (see the expressions in eqns. 5.38 and 5.39):

$$P_{ij} = \int_{v_1}^{v_2} \int_{u_1}^{u_2} u^{i-1} v^{j-1} g(R)\, \mathrm{d}u\, \mathrm{d}v \tag{A2.1}$$

and

$$Q_{ij} = \int_{v_1}^{v_2} \int_{u_1}^{u_2} u^{i-1} v^{j-1} \frac{1}{R} \frac{\mathrm{d}g(R)}{\mathrm{d}R}\, \mathrm{d}u\, \mathrm{d}v \tag{A2.2}$$

The Green function $g(R)$ and the distance R are evaluated as

$$g(R) = \frac{\exp(-j\beta R)}{4\pi R} \tag{A2.3}$$

$$R = |\boldsymbol{r}'(u, v) - \boldsymbol{r}| \tag{A2.4}$$

where $\boldsymbol{r}'(u, v)$ is the parametric equation of the bilinear surface:

$$\boldsymbol{r}'(u, v) = \boldsymbol{r}_c + \boldsymbol{r}_u u + \boldsymbol{r}_v v + \boldsymbol{r}_{uv} uv \tag{A2.5}$$

By substituting the expression in eqn. A2.5 into eqn. A2.4, after simple rearrangements the distance R can be written in the form

$$R = \sqrt{[e_u\{(u - u_0)^2 + d_u^2\}]} \tag{A2.6}$$

where the Lamé coefficient e_u and the coefficients u_0 and d_u are obtained as

$$e_u = |\boldsymbol{r}_u + \boldsymbol{r}_{uv} v| \tag{A2.7}$$

$$\left.\begin{aligned} u_0 &= \frac{(\boldsymbol{r} - \boldsymbol{r}_c - \boldsymbol{r}_v v).(\boldsymbol{r}_u + \boldsymbol{r}_{uv} v)}{e_u^2} \\ d_u^2 &= \frac{(\boldsymbol{r} - \boldsymbol{r}_c - \boldsymbol{r}_v v)^2}{e_u^2} - u_0^2 \end{aligned}\right\} \tag{A2.8}$$

Note that the expression in eqn. A2.6 has the same form as that in eqn. A1.8.

Therefore the integrals P_{ij} and Q_{ij} can be expressed in terms of the integrals P_i and Q_i as

$$P_{ij} = \int_{v_1}^{v_2} v^{j-1} P_i \, dv \tag{A2.9}$$

$$Q_{ij} = \int_{v_1}^{v_2} v^{j-1} Q_i \, dv \tag{A2.10}$$

The problem of the first integration in the integrals P_{ij} and Q_{ij} has thereby been solved. With this first integration a smoother integrand is obtained. Nevertheless, it is also necessary to pay considerable attention to the second integration.

It is obvious that when the field point is along the segment of integration of the integrals P_i and Q_i, these integrals have infinite values. (Note that this is never the case with wires, but only with bilinear surfaces.) We shall therefore consider only that case.

Let us concentrate our attention first on the integrals of the P_{ij} type. Starting from eqn. A2.1, it is possible to construct integrals P'_{ij} of the form

$$P'_{ij} = \int_{v_1}^{v_2} \int_{u_1}^{u_2} \left(u^{i-1} v^{j-1} g(R) - u_{00}^{i-1} v_{00}^{j-1} \frac{1}{4\pi R} \right) du \, dv \tag{A2.11}$$

where u_{00} and v_{00} are the u and v co-ordinates of the point at which the potential is calculated. (This is a singular point of the integral P_{ij}.) The integrand of the integral P'_{ij} obviously does not contain singular points. Bearing in mind the procedure used for obtaining the expression in eqn. A2.9, the integral P'_{ij} can be written as

$$P'_{ij} = \int_{v_1}^{v_2} \left(v^{j-1} P_i - \frac{u_{00}^{i-1} v_{00}^{j-1}}{4\pi} \int_{u_1}^{u_2} \frac{1}{R} du \right) dv \tag{A2.12}$$

Note that the integrand of this integral also contains no singular points. With the normalisation given in eqn. A1.11, the integral in parentheses can be expressed as

$$\int_{u_1}^{u_2} \frac{1}{R} du = \frac{1}{e_u} \int_{s_1}^{s_2} \frac{1}{R_0} ds = \frac{1}{e_u} \ln \frac{(s_2 - s_0 + R_0)}{(s_1 - s_0 + R_0)} \tag{A2.13}$$

Substituting this expression into eqn. A2.12 and taking account of eqn. A1.16, after simple rearrangements the integral P'_{ij} can be written in the form

$$P'_{ij} = P''_{ij} - \frac{u_{00}^{i-1} v_{00}^{j-1}}{4\pi} \int_{v_1}^{v_2} \frac{1}{e_u} \ln\{(s_2 - s_0 + R_0)(s_0 - s_1 + R_0)\} \, dv \tag{A2.14}$$

where

$$P''_{ij} = \int_{v_1}^{v_2} \left(v^{j-1} P_i + \frac{u_{00}^{i-1} v_{00}^{j-1}}{4\pi e_u} \ln(d_s^2) \right) dv \tag{A2.15}$$

The integrand in the integral in eqn. A2.14 has no singular points. We have already seen that the same is valid for the integrand of the integral P'_{ij}. Therefore the integrand of the integral P''_{ij} has no singular points either. This conclusion remains valid if in the integral P''_{ij} the function $e_u(v)$ is replaced by its value at the singular point $e_u(v_{00})$. In that case the integral P''_{ij} takes the form

$$P''_{ij} = \int_{v_1}^{v_2} \left(v^{j-1} P_i + \frac{u_{00}^{i-1} v_{00}^{j-1}}{4\pi e_a(v_{00})} \ln(d_s^2) \right) dv \qquad \text{(A2.16)}$$

Now instead of using eqn. A2.9, the integral P_{ij} can be evaluated as

$$P_{ij} = P''_{ij} - \frac{u_{00}^{i-1} v_{00}^{j-1}}{4\pi e_u(v_{00})} \int_{v_1}^{v_2} \ln(d_s^2)\, dv \qquad \text{(2.17)}$$

where P''_{ij} is evaluated numerically on the basis of eqn. A2.16 and the rest is evaluated analytically.

This method of evaluating the integrals should also be used if the point is close to the bilinear surface, and not only when it is on it. A similar process is used to evaluate the integrals Q_{ij}. However, with the integrals Q_{ij} we have additional problems. For a field point at the bilinear surface, in the general case these integrals have infinite values. In numerical evaluation of the electric and magnetic field vectors, only appropriate combinations of these integrals yield finite values which correspond to the tangential component of the electric-field vector and the normal component of the magnetic-field vector. This problem can be circumvented in several ways. The simplest is probably to position the field point close to the bilinear surface, on either side of it, instead of actually at the surface.

Index

www.ingramcontent.com/pod-product-compliance
Lightning Source LLC
Chambersburg PA
CBHW031951180326
41458CB00006B/1692